T0214280

Communications in Computer and Information Science 1197

Commenced Publication in 2007
Founding and Former Series Editors:
Phoebe Chen, Alfredo Cuzzocrea, Xiaoyong Du, Orhun Kara, Ting Liu,
Krishna M. Sivalingam, Dominik Ślęzak, Takashi Washio, Xiaokang Yang,
and Junsong Yuan

More information about this series at http://www.springer.com/series/7899

Giorgos Flouris · Dominique Laurent ·
Dimitris Plexousakis · Nicolas Spyratos ·
Yuzuru Tanaka (Eds.)

Information Search, Integration, and Personalization

13th International Workshop, ISIP 2019
Heraklion, Greece, May 9–10, 2019
Revised Selected Papers

 Springer

Editors
Giorgos Flouris (iD)
Foundation for Research
and Technology Hellas
Heraklion, Greece

Dimitris Plexousakis
Foundation for Research
and Technology Hellas
Heraklion, Greece

Yuzuru Tanaka
Hokkaido University
Sapporo, Japan

Dominique Laurent (iD)
University of Cergy-Pontoise
Cergy Pontoise, France

Nicolas Spyratos
University of Paris-Sud
Orsay, France

ISSN 1865-0929   ISSN 1865-0937   (electronic)
Communications in Computer and Information Science
ISBN 978-3-030-44899-8   ISBN 978-3-030-44900-1   (eBook)
https://doi.org/10.1007/978-3-030-44900-1

This Springer imprint is published by the registered company Springer Nature Switzerland AG
The registered company address is: Gewerbestrasse 11, 6330 Cham, Switzerland

Preface

This book contains the selected research papers presented at ISIP 2019, the 13th International Workshop on Information Search, Integration and Personalization, during May 9–10, 2019, at FORTH Institute of Computer Science, Heraklion, Greece. Two keynote talks were given during the workshop:

- "Towards diversity-aware, fair and unbiased data management," by Professor Evaggelia Pitoura, University of Ioannina, Greece.
- "Visual analytics of multiple media and real world Big Data," by Professor Masashi Toyoda, University of Tokyo, Japan.

There were 24 presentations of scientific papers, of which 16 were submitted to the post-workshop peer review. The International Program Committee selected 11 papers to be included in the proceedings.

The themes of the presented and/or submitted papers reflected today's diversity of research topics as well as the rapid development of interdisciplinary research. With increasingly sophisticated research in science and technology, there is a growing need for interdisciplinary and international availability as well as distribution and exchange of the latest research results in organic forms, including not only research papers and multimedia documents, but also various tools developed for measurement, analysis, inference, design, planning, simulation, and production as well as the related large data sets. Similar needs are also growing for the interdisciplinary and international availability as well as distribution and exchange of ideas of works among artists, musicians, designers, architects, directors, and producers. These contents, including multimedia documents, application tools, and services are being accumulated on the Web, as well as in local and global databases, in a remarkable speed that we have never experienced with other kinds of publishing media. Large amounts of content are now already on the Web, waiting for their advanced personal and/or public reuse. We need new theories and technologies for the advanced information search, integration through interoperation, and personalization of Web content as well as database content.

The ISIP 2019 workshop was organized to offer a forum for presenting original work and stimulating discussions and exchanges of ideas around these themes, focusing on the following topics.

- Data Analytics and Visualization
- Data mining
- Linked/Open Data
- Languages and Query Answering
- Data Integration, Data Warehouses, and Data Lakes
- Gamification and Recommendation
- Machine Learning

The selected papers contained in this book are grouped into four major topics, namely Linked Data, Data Analytics, Data Integration, and Data Mining Applications; they span major current topics in Information Management research.

Historical Note

ISIP started as a series of Franco-Japanese workshops in 2003, and its first edition was placed under the auspices of the French embassy in Tokyo, which provided the financial support along with JSPS (Japanese Society for the Promotion of Science). Up until 2012, the workshops have alternated between Japan and France, and attracted increasing interest from both countries. Then, motivated by the success of the first editions of the workshop, participants from countries other than France or Japan volunteered to organize it in their home country. The following shows the history of past ISIP workshops:

- 2003: 1st ISIP in Sapporo (June 30 – July 2, Meme Media Lab, Hokkaido University, Japan)
- 2005: 2nd ISIP in Lyon (May 9–11, University Lyon 1, France)
- 2007: 3rd ISIP in Sapporo (June 27–30, Meme Media Laboratory, Hokkaido University, Japan)
- 2008: 4th ISIP in Paris (October 6–8, Tour Montparnasse, Paris, France)
- 2009: 5th ISIP in Sapporo (July 6–8, Meme Media Laboratory, Hokkaido University, Japan)
- 2010: 6th ISIP in Lyon (October 11–13, University Lyon 1, France)
- 2012: 7th ISIP in Sapporo (October 11–13, Meme Media Laboratory, Hokkaido University, Japan)
- 2013: 8th ISIP in Bangkok (September 16–18, Centara Grand & Bangkok Convention Centre CentralWorld Bangkok, Thailand)
- 2014: 9th ISIP in Kuala Lumpur (October 9–10, HELP University, Kuala Lumpur, Malaysia)
- 2015: 10th ISIP in Grand Forks (October 1–2, University of North Dakota, Grand Forks, North Dakota, USA)
- 2016: 11th ISIP in Lyon (November 3–4, University Lyon 1, France)
- 2018: 12th ISIP in Kyushu (May 14–15, Kyushu University, Fukuoka, Japan)

Originally, the workshops were intended for a Franco-Japanese audience, with the occasional invitation of researchers from other countries as keynote speakers. The proceedings of each workshop were published informally, as a technical report of the hosting institution. One exception was the 2005 workshop, selected papers of which were published by the *Journal of Intelligent Information Systems* in its special issue for ISIP 2005 (Vol. 31, Number 2, October 2008). The original goal of the ISIP workshop series was to create close synergies between a selected group of researchers from the two countries; and indeed, several collaborations, joint publications, joint student supervisions, and research projects originated from participants of the workshop.

After the first six workshops, the organizers concluded that the workshop series had reached a mature state with an increasing number of researchers participating every year. As a result, the organizers decided to open up the workshop to a larger audience by inviting speakers from over ten countries at ISIP 2012, ISIP 2013, ISIP 2014, as well as at ISIP 2015. The effort to attract an even larger international audience has led to organizing the workshop in countries other than France and Japan. This will continue in the years to come. Especially in these past four years, an extensive effort was made to include in the Program Committee academics coming from around the globe, giving the workshop an even more international character.

We would like to express our appreciation to all the staff members of the organizing institution for the help, kindness, and support before, during, and after the workshop. Of course we also would like to cordially thank all speakers and participants of ISIP 2019 for their intensive discussions and exchange of new ideas. This book is an outcome of those discussions and exchanged ideas. Our thanks also go to the Program Committee members whose work has been undoubtedly essential for the selection of the papers contained in this book.

January 2020

Dimitris Plexousakis
Nicolas Spyratos
Yuzuru Tanaka

Organization

Executive Committee

Co-chairs

Dimitris Plexousakis	FORTH-ICS, Greece
Nicolas Spyratos	Paris-Sud University, France
Yuzuru Tanaka	Hokkaido University, Japan

Program Committee Chairs

Giorgos Flouris	FORTH-ICS, Greece
Dominique Laurent	University of Cergy-Pontoise, France

Local Organization

Haridimos Kondylakis	FORTH-ICS, Greece

Publicity Chair

Ioannis Chrysakis	FORTH-ICS, Greece

Program Committee

Antonis Bikakis	University College London, UK
Yeow Wei Choong	HELP University, Malaysia
Ioannis Chrysakis	Ghent University, Belgium, and FORTH-ICS, Greece
Giorgos Flouris	FORTH-ICS, Greece
Arnaud Giacometti	Université François Rabelais de Tours, France
Mirian Halfeld Ferrari	Université d'Orléans, France
Tao-Yuan Jen	University of Cergy-Pontoise, France
Haridimos Kondilakis	FORTH-ICS, Greece
Dimitris Kotzinos	University of Cergy-Pontoise, France
Anne Laurent	Université Montpellier, France
Dominique Laurent	University of Cergy-Pontoise, France
Yoshihbiro Okada	Kyushu University, Japan
Laurent d'Orazio	Université de Rennes 1, France
George Papastefanatos	Institute for the Management of Information Systems, Greece
Jean-Marc Petit	INSA de Lyon, France
Dimitris Plexousakis	FORTH-ICS, Greece
Pascal Poncelet	Université Montpellier, France
Lakhdar Sais	Université d'Artois, France
Domenico Fabio Savo	University of Bergamo, Italy

Contents

Linked Data

Enabling Efficient Question Answering over Hundreds of Linked Datasets

Eleftherios Dimitrakis[1,2], Konstantinos Sgontzos[1,2],
Michalis Mountantonakis[1,2(✉)], and Yannis Tzitzikas[1,2]

[1] Institute of Computer Science, FORTH, Heraklion, Greece
{dimitrakis,sgontzos,mountant,tzitzik}@ics.forth.gr
[2] Computer Science Department, University of Crete, Heraklion, Greece

Abstract. In this paper we introduce an approach, called `LODQA`, for open domain Question Answering over Linked Open Data. We confine ourselves to three kinds of questions: factoid, confirmation, and definition questions. By using `LODQA` it is feasible to answer questions over 400 millions of entities of any domain without using any training data, since we exploit simultaneously 400 Linked datasets. In particular, we exploit the services of `LODsyndesis`, a suite of services (based on semantics-aware indexes) which supports cross-dataset reasoning over hundreds of Linked datasets and 2 billion triples. The proposed Question Answering process follows an information extraction approach and comprises several steps including question cleaning, heuristic based question type identification, entity recognition, linking and disambiguation using Linked Data-based methods and pure NLP methods (specifically DBpedia Spotlight and Stanford CoreNLP), WordNet-based question expansion for tackling the lexical gap (between the input question and the underlying sources), and triple scoring for producing the final answer. We discuss the benefits of this approach in terms of answerable questions and answer verification, and we investigate, through experimental results, how the aforementioned steps of the process affect the effectiveness and the efficiency of question answering.

Keywords: Questions Answering · Linked data · Multiple datasets

1 Introduction

Although the first QA (Question Answering) systems were created decades ago (back in 1960s), the problem is still open since the existing techniques have several limitations (for more see [24]), therefore QA is subject of continuous research. There is a wide range of techniques for QA ranging from simple manually-written regular expression-based methods, to methods relying on deep learning, e.g. see the survey papers [16,21,28], and there are several collections for evaluating QA systems (see [8]). Recently we observe a wide adoption of QA-based personal assistants (including Apple's Siri, Google Assistant, Amazon's Alexa) that are capable of answering a wide range of questions, as

© Springer Nature Switzerland AG 2020
G. Flouris et al. (Eds.): ISIP 2019, CCIS 1197, pp. 3–17, 2020.
https://doi.org/10.1007/978-3-030-44900-1_1

well as an increasing interest from the database community for natural language interfaces to databases [2, 26]. Indeed natural language interfaces can complement the existing methods for query formulation by casual users, i.e. faceted search [27], as evidenced by prototypes supporting spoken dialogue interfaces for information navigation [20].

Open domain (as opposed to closed domain) Question Answering (QA) is a challenging task, since it requires to tackle a number of issues: (i) the issue of data distribution, i.e., several datasets, that are usually distributed in different places, should be exploited for being able to support open domain question answering, (ii) the difficulty of word sense disambiguation, because the associated vocabulary is not restricted to a single domain, and (iii) the difficulty (or inability) to apply computationally expensive techniques, such as deep NLP analysis, due to the huge size of the underlying sources. In this paper we focus on Open Domain Question Answering over *Linked Data*. We introduce LODQA, a Linked Data-based Question Answering system that exploits LODsyndesis [17], a recently launched suite of services over hundreds of LOD Datasets (that contains two billion triples about 400 million entities). We selected to use LODsyndesis, because if offers two distinctive features for the QA process, which are not supported by a single source: (a) it is feasible to verify an answer to a given question from several sources, and (b) the number of questions that can be answered is highly increased, because datasets usually contain complementary information for the same topics and entities. Essentially, we try to find the best triple(s) for answering the incoming question; we do not carry out any other information integration techniques (like those surveyed in [19]).

Regarding (a), suppose that the given question is *"What is the population of Heraklion?"*, and the system retrieves two candidate triples, i.e., {(Heraklion, population, 140,730), (Heraklion, population, 135,200)}. The two triples contain a different value for the population of that city, however, suppose that the first triple can be verified from four datasets (say D_1, D_2, D_3, D_4), whereas the second one only from a single dataset (say D_5). In this example, LODQA will return as correct answer the first triple, because it can be verified from a larger number of datasets, thereby, we have more evidence about its correctness.

Regarding (b), suppose that LODQA receives the two following questions for an other domain (say marine domain), *"Is Yellowfin Tuna a predator of Atlantic pomfret?"* and *"Which is the genus of Yellowfin Tuna?"*. These two questions are addressed to the same entity i.e. *"Yellowfin Tuna"*, however there is not a single dataset where we can find the desired information for answering both questions. Indeed, LODQA is able to answer the first question by using data from *Ecoscope* dataset, whereas the second question is answerable by a triple that occurs in *DBpedia* knowledge base.

Concerning the Question Answering process followed from LODQA, it is an information extraction approach, as opposed to the semantic parsing approach, that consists of multiple steps. In particular, LODQA performs question cleaning (e.g., removal of stopwords) and it identifies the question type (e.g., factoid) by exploiting heuristics. Moreover, it recognizes the entities of the question and it performs linking and disambiguation, by using both pure NLP methods and Linked Data-based methods, specifically Stanford CoreNLP [9,13] and DBpedia Spotlight [14]. Furthermore, it uses *WordNet* [15] for tackling the possible lexical gap between a given question and the answer which can be found in the underlying sources. Finally, it receives the candidate triples from LODsyndesis, and it scores each candidate triple for producing the final answer. Concerning evaluation, we discuss the benefits of this approach in terms of answerable questions and answer verification, and we investigate through experimental results, how the aforementioned steps of the process affect the effectiveness and the efficiency of question answering.

The rest of this paper is organized as follows: Sect. 2 discusses related work, Sect. 3 introduces the proposed approach, Sect. 4 reports comparative experimental results, whereas Sect. 5 describes an application of the proposed approach. Finally, Sect. 6 concludes the paper and discusses directions for future research and work.

2   Related Work

Knowledge Base Question Answering (KBQA) systems can be divided in two different categories: (a) *Semantic Parsing (SP)* [4,10,23,30,31], and (b) *Information Extraction (IE)* [1,3,7,22,29].

Concerning *SP approaches*, they focus on question understanding, i.e., they convert sentences into their semantic representation and they usually generate a query (e.g., a SPARQL query), for retrieving the answer. Such approaches can answer compositional questions by using aggregation operators (e.g., argmax, count), however, they suffer from structure differences between the Knowledge Base and the input Natural Language question. On the contrary, the objective of *IE approaches* is to identify the main entities of the question and to map the words of the question to the Knowledge base predicates, either by using pre-defined templates, or automatically generated ones. As a final step, these approaches exploit the neighborhood (in the knowledge graph) of each matched entity for producing the final answer. Their disadvantage is that they cannot easily answer compositional questions, since they cannot represent such operators [11]. Our work, i.e., LODQA, belongs to IE category.

The most related approaches to LODQA, are predominantly *WDAqua* [7] and *AMAL* [22], and secondarily Aqqu [3,32] and SINA [25]. In contrast to these four related tools, LODQA exploits the contents of 400 datasets for answering a given question, whereas the other tools support either a single or a few KBs, therefore, they cannot verify the answers from several datasets. LODQA follows an *information extraction approach* by exploiting the services and indexes of LODsyndesis, instead of using a SPARQL translation approach. By using indexes, we can offer faster question responses comparing to approaches using SPARQL queries, since their efficiency usually rely on the sources' servers, whereas SPARQL querying can be quite expensive for large knowledge bases.

Comparing to WDAqua [7], we take into account both the syntactic form of the question and the relations of the underlying question words, instead of exploiting only the semantics of the question words. However, we do not support multilingual questions. Concerning the differences with AMAL [22], the latter exploits Wikipedia Disambiguation links and DBpedia lexicons for performing relation matching, whereas we use services offered by LODsyndesis for taking into consideration equivalent relationships, and also synonyms through WordNet. Therefore, we can exploit multiple sources for relation matching task, instead of using only DBpedia resources. On the contrary, we do not support list and aggregation questions, which are offered from AMAL [22]. Regarding Aqqu [3], we exploit two different tools for entity detection, i.e., DBpedia Spotlight and Stanford CoreNLP, whereas Aqqu [3] uses hand-crafted rules based on POS-tags. Concerning SINA [25], it performs the data interlinking among the datasets (for a few number of datasets) at query time (which can be time-consuming), whereas, we exploit the indexes of LODsyndesis, where the interlinking has already been done once at indexing time. Finally, comparing to Aqqu [3] and SINA [25], we do not use any training data for producing the answer.

3   The LODQA Approach

In Sect. 3.1 we present the LODsyndesis services that we exploit, while in Sect. 3.2, we introduce the proposed QA process.

3.1   LODsyndesis Knowledge Services

We decided to use LODsyndesis, for two tasks: namely *Entity Detection* and *Answer Extraction*, due to the following benefits that cannot be found in a single knowledge base: (a) it collects all the available information for millions of entities from hundreds of datasets, (b) it contains complementary information from different datasets. Moreover, it is worth mentioning that (c) it can surpass the problems of non-informative URIs, since it supports cross-dataset reasoning. The process of indexing of LODsyndesis is illustrated in Fig. 1, i.e., LODsyndesis uses as input several datasets containing RDF triples (see the lower left side of Fig. 1), where a triple is a statement of the form subject-predicate-object (s,p,o) and T is the set of all the triples that exist in the LOD cloud. Moreover, it uses several equivalence relationships (see the lower right side of Fig. 1), such as owl:sameAs relationships which denote that two URIs refer to the same entity (e.g., dbp:Heraklion ≡ test:Heraklion), and owl:equivalentProperty relationships which are used for denoting that two schema elements are equivalent (e.g., dbp:population ≡ test:population). LODsyndesis uses as input these equivalence relationships and computes their transitive and symmetric closure for collecting all the information for any entity (e.g., see the index for "Heraklion" in the middle part of Fig. 1).

Concerning benefit (a), it is important for any kind of question to verify the answer from several sources. Regarding benefit (b), for any type of question, two or more datasets can possibly answer different questions, e.g., in Fig. 1,

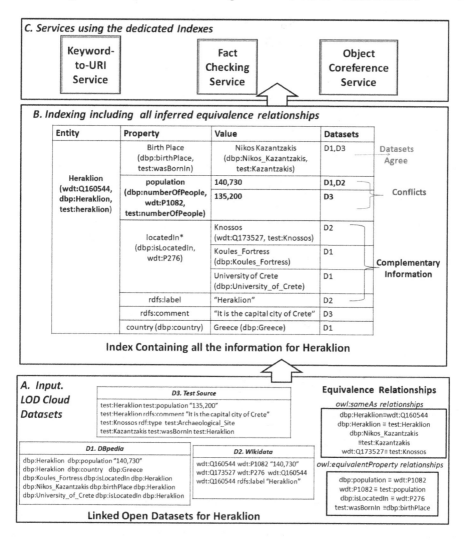

Fig. 1. The steps of LODsyndesis

one dataset contains a comment about Heraklion, another one about the country where Heraklion is located in, and so forth. Concerning benefit (c), many datasets publish non-informative URIs, e.g., in Fig. 1 only Wikidata can answer the question "Is Knossos located in Heraklion?", with the corresponding triple (wdt:Q173527, wdt:P276, wdt:Q160544). LODsyndesis supports cross-dataset reasoning, i.e., it computes the transitive and symmetric closure of equivalent relationships (e.g., see the lower right side of Fig. 1) and it stores the equivalent URIs of each URI, thereby, it knows that dbp:Heraklion ≡ wdt:Q160544, dbp:isLocatedIn ≡ wdt:P276 and wdt:Q173527 ≡ test:Knossos. Therefore, we can find fast the correct answer, by checking the equivalent URIs of each one.

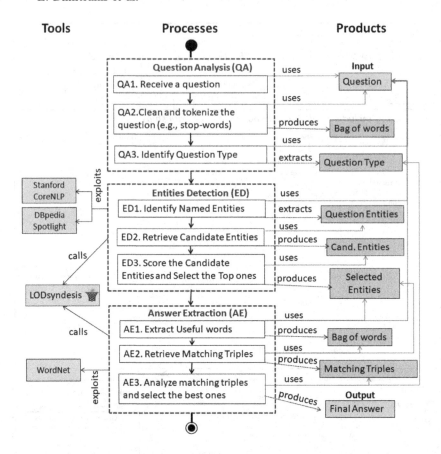

Fig. 2. Overview of the QA process

LODsyndesis offers several services by exploiting the aforementioned semantics-aware indexes. Concerning LODQA, it exploits the following three LODsyndesis services (more information about them can be found in [18]): (a) the *Keyword-to-URI* service which returns those URIs whose suffix starts with a given keyword, (b) the *Object Coreference* service which provides all the equivalent URIs for a given one, and (c) the *Fact Checking* service, which can be used for retrieving all the triples containing a set of given keywords for a single focused entity.

3.2   The Process of LODQA

The question answering process consists of three main phases: *(i) Question Analysis (QA), (ii) Entities Detection (ED)* and *(iii) Answer Extraction (AE)*. Figure 2 introduces the main phases and steps of LODQA process, where one can clearly observe the input and the output of each different step. To make these steps more clear, Fig. 3 shows a running example i.e., the steps for answering the factoid question "What is the population of Heraklion?".

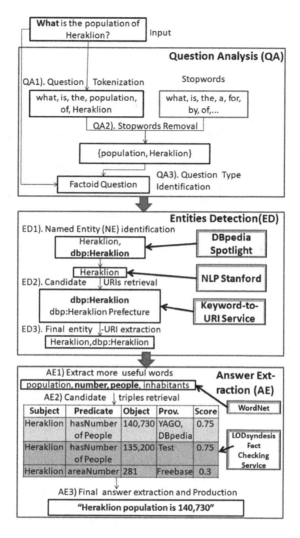

Fig. 3. The QA process over the running example

Question Analysis **Phase.** In this phase, we convert the input question into a set of tokens and we remove the stopwords of the given question, such as the words "what" and "is" in the example of Fig. 3. The next step is the question type identification, by using a set of indicative words and simple heuristics. Concerning factoid and confirmation questions, we check if the question starts with one of the following words: $W_{factoid} = \{$when, who, where, what, which, ...$\}$ for factoid questions, and $W_{confirm} = \{$are, did, is, was, does, were, do, ...$\}$ for confirmation questions. Finally, for the definition questions, we check if the question contains one of the following words: $W_{def} = \{$mean, meaning, definition, ...$\}$ in its middle part. As an example, in Fig. 3 we identified the question as a factoid one, since is starts with the word "what".

***Entities Detection* Phase.** The target of the second phase is to identify the entities occurring in the question and to link them with their corresponding URIs in the sources which are supported from LODQA. For achieving this goal, we exploit two widely used tools, i.e., Stanford CoreNLP [9,13] and DBpedia Spotlight [14]. The Stanford CoreNLP tool, hereafter *SCNLP*, combines hand-crafted rules and statistical sequence taggers for identifying the named entities of a given question, and it returns the recognized entities in natural language, e.g., for the input question of Fig. 3, it will return as the entity of the question the word "Heraklion". However, since LODQA needs also the corresponding URI of each entity, we use the *Keyword-to-URI* service of LODsyndesis, which returns a set of candidate URIs for a given keyword. On the contrary, DBpedia Spotlight uses a string matching algorithm, a lexicon for retrieving the possible candidates and a TF*IDF variation for tackling disambiguation issues. As an output, it produces for each entity a pair containing the entity in natural language and its corresponding URI in *DBpedia* knowledge base [12]. Afterwards, we compare the candidate URIs derived from both tools for the same entity, we compute a score for each such URI, and we select the most relevant URI for each entity. In our running example of Fig. 3, we identified that for the entity "Heraklion", the most relevant URI is "dbp:Heraklion" (and not "dbp:Heraklion_Prefecture").

***Answer Extraction* Phase.** In the third phase, the objective is to retrieve the candidate RDF triples, and to identify the best matching triple, for returning the final answer. It is achieved through the exploitation of LODsyndesis, and of an *expanded set of question words* by using (i) the SCNLP lemmatizer and (ii) the WordNet dictionary [15]. Concerning SCNLP lemmatizer, we use it for extracting the lemmas of the question words, e.g., the lemma of the word "analyzed" is "analyze". Regarding WordNet dictionary, we use the API offered by extJWNL[1], for deriving nouns, verbs and synonyms based on the POS tags of the question words. Therefore, it can produce from the word "populated" the noun "population", and from the word "population", the word "inhabitants" and the phrase "number of people" which have a similar meaning.

The next step is to exploit the *factChecking* service of LODsyndesis, which takes as input a single entity, along with a set of words, i.e., in our running example, we give as input to that service the parameters "dbp:Heraklion" and the words "population", "inhabitants" and "number of people" (e.g., the latter two phrases derived through WordNet dictionary). Afterwards, a set of candidate triples is returned from LODsyndesis, which are analyzed through LODQA for selecting the most relevant answer for the given question. In our running example, we received three candidate triples for the input question. However, it is worth mentioning that without the *Question Expansion* step, it would be infeasible to derive candidate triples for the given question (i.e., there was not a triple containing the word "population" in this example).

Afterwards, LODQA produces the final triple for the input question, by taking into account its type. Specifically, for factoid questions, it selects the max scored triple based on the percentage of the question words included in the triple,

---

[1] https://github.com/extjwnl/extjwnl.

and the number of provenance datasets. In our running example, by analyzing the candidate triples, we identified that the first two triples were more relevant comparing to the third one. However, LODQA selected the first triple (i.e., the population of "Heraklion" is "140,730") as the best matching one, since it was included in more datasets in comparison to the second triple. Concerning confirmation questions, LODQA returns "Yes" if the candidate triple contains all the entities of the question and at least one other "useful" word, e.g., suppose that the given question is the following: "Was Nikos Kazantzakis born in Heraklion?" and the candidate triple is "Nikos Kazantzakis, birthPlace, Heraklion", the system would return "Yes", since the answer contains both entities and the predicate "birthPlace", which is a synonym to the predicate "born in". Finally, for definition questions, it returns the best matching triple containing as predicate one of the following: *rdfs:comment, dcterms:description, dbpedia:abstract*.

4   Evaluation

Here, in Sects. 4.1–4.4 we report experimental results concerning the effectiveness and efficiency of *Entities Detection* and *Answer Extraction* steps, by using SimpleQuestions (v2) collection [6][2]. Finally, in Sect. 4.5, we show some measurements regrading the impact of using multiple datasets.

4.1   Evaluation Collection

We performed a comparative evaluation over the SimpleQuestions (v2) collection [6], for evaluating and improving the tasks of *Entities Detection* and *Answer Extraction*. This collection contains 108,442 simple (mainly factoid) questions, i.e., questions that can be answered by using a single triple from Freebase knowledge base [5]. For each question it includes the corresponding answer, i.e., a single Freebase triple. It is worth mentioning that LODsyndesis contains information from several sources (including DBpedia, Freebase and others), therefore LODQA can answer a question by exploiting a different dataset (e.g., DBpedia) and not Freebase. Since this requires a manual check for evaluating whether the answer is correct, mainly due to missing mappings between these sources (e.g., between DBpedia and Freebase), we selected a subset of them for the experiments. Indeed, we selected randomly a set 1,000 factoid questions, where each question contained on average 7 words. The subset of the collection that was used in the experiments, is accessible online[3].

4.2   Entities Detection Evaluation

Our objective is to understand how the capabilities of the two used different tools (SCNLP and DBpedia Spotlight) affect the outcome of the whole process.

---

[2] http://research.fb.com/downloads/babi/.

[3] http://islcatalog.ics.forth.gr/tr/dataset/simplequestions-v2-1000-questions.

Table 1. Evaluation using 1000 questions from SimpleQuestions (v2). Left: Accuracy of each *Named Entity Recognition* approach. Right: Accuracy of each *Triples Retrieval* approach

NER Method	Accuracy		Model	Accuracy	Accuracy (Perfect ED)
SCNLP-Spotlight	0.626		LODQA	**0.487**	**0.642**
Spotlight-SCNLP	0.653		LODQA-w/o-L	0.411	0.556
Combined	**0.737**		LODQA-w/o-N	0.414	0.558
			LODQA-w/o-V	0.429	0.581
			LODQA-w/o-LNV	0.407	0.547

For this reason, we report comparative results by using three different approaches. Specifically, for each approach we measure the accuracy, i.e., the number of questions where each approach identified the correct entities, divided by the number of all questions. The approaches which are compared for the Named Entity (NE) detection and linking follow: (i) *SCNLP-Spotlight*: we use SCNLP and in case of failing to recognize any NE, we use DBpedia Spotlight, (ii) *Spotlight-SCNLP*: we use DBpedia Spotlight and in case of failure, we use SCNLP, (iii) *Combined*: we exploit both tools for identifying the NEs and their URIs and then, we use some simple heuristics for selecting the best entities. The evaluation results are shown in Table 1(left). We observe that the combined method achieves much higher accuracy, i.e. 0.73, compared to any of the other two approaches, which achieve an accuracy of 0.62 and 0.65, respectively! For this reason, we will use that method for evaluating the outcome of the whole process in Sect. 4.3.

4.3   Answer Extraction Evaluation

Regarding the *Answer Extraction* step, our target is to tackle the possible *lexical gap* between the question and the underlying datasets. For this reason, we compare variations of our approach, where we expand the available set of questions words. Indeed, we perform the expansion by producing the lemmas (from SCNLP) of the question words. Moreover, based on the POS tag of each word, if (a) a word is a *Verb*, we produce all the derived nouns (from WordNet) and (b) if it is a *Noun*, we produce all the derived verbs. We evaluate the effectiveness of our approach (i.e., LODQA) by using all the aforementioned expansion methods (i.e., lemmas, nouns, verbs), and we compare it with variations of our approach that do not perform word expansion based on lemmas (LODQA-w/o-L), nouns (LODQA-w/o-N), and verbs (LODQA-w/o-V). Moreover, we provide also experimental results for an approach that do not perform any word expansion, i.e., (LODQA-w/o-LNV). The evaluation results are shown in Table 1(Right), where we measure the accuracy of each different variation, i.e., the number of questions answered correctly, divided by the number of all the questions. The proposed approach (i.e., LODQA) using all the expansion steps achieves the highest accuracy

Table 2. Efficiency results using 500 questions from SimpleQuestions (v2).

Step	Average time
Question analysis	0.007 s
Entities detection	1.808 s
Query expansion	0.330 s
Candidate triples retrieval	3.005 s
Final answer production	0.134 s
Total time	5.330 s

Table 3. LODsyndesis measurements

Measurement	Value
Number of entities in ≥ 2 datasets	25,289,605
Number of entities in ≥ 3 datasets	6,979,109
Verifiable questions from at least 2 datasets	28,439,760
Average triple per entity (by using 1 dataset)	17.3
Average triple per entity (by using all datasets)	29.3

(i.e., 0.49), whereas by taking into account only the questions that passes the Entities Detection Step (i.e., all the questions that we detected the correct entities), the accuracy increases (i.e., 0.64). On the contrary, without any question words expansion, the precision is only 0.4 and 0.54 respectively. Concerning the different methods for expanding the set of question words, we identified that for this set of questions, verbs were more important that nouns and lemmas. The above evaluation results indicate that our approach is KB agnostic, since it can be applied for any given KB (indexed by LODsyndesis) without requiring any additional effort and training data.

4.4   Efficiency

For performing the experiments, we used a single machine with 8 GB RAM, 8 cores and 60 GB Disk space, and we measured the efficiency in 500 questions of SimpleQuestions collection for each different step, as it can be seen in Table 2. As we can see, LODQA needs on average 5.33 s to answer a question. The most time consuming steps are to retrieve the candidate triples (57.2% of the required time) and to detect the entities of the question (30% of the required time). Furthermore, it is worth noting that the minimum time for answering a question was 1.6 s and the maximum one was 37.46. Finally, half of the questions (i.e., median value) were answered in less than 3.7 s.

4.5   The Benefits of Using Multiple Datasets

Table 3 shows measurements for evaluating the impact of using multiple datasets (and of performing cross-dataset reasoning). Particularly, LODsyndesis contains information for 25.2 million entities from at least two datasets, whereas for 6.9 million entities we can retrieve information from at least three datasets. It is worth noting that there exists 28.4 million of possible questions that can be verified by more than one dataset, i.e., corresponds to simple questions answerable from at least two datasets. Therefore, it is evident that by using multiple datasets, we increase the probability of answering a given question, whereas the number of verifiable questions is increased, too. Moreover, for these 25.2 million entities, if we use only a single dataset (even the dataset containing the most triples for each entity), the average number of triples per entity is 17.3. On the contrary, due to the cross-dataset reasoning (i.e., computation of transitive and symmetric closure of equivalent relationships), `LODsyndesis` collects all the available information for each entity from all the datasets. Due to this process, the average number of triples of each of these entities highly increases (i.e., it becomes 29.3).

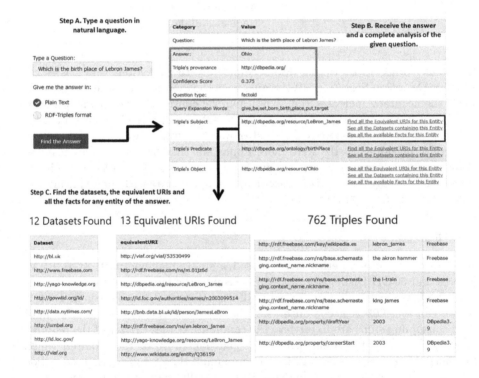

Fig. 4. An example of the LODQA demo

5 Web Demo and Related Links

The `LODQA` is currently hosted and runs in a single machine of *okeanos* cloud computing service (https://okeanos.grnet.gr/) with an *i5* core, 8 GB main memory and 60 GB disk space. Although the hosting machine has a low computational power, the interaction is real time, i.e., few seconds are needed to answer a question.

In the *website* https://demos.isl.ics.forth.gr/LODQA/, we offer a list of demo factoid, confirmation and definition questions, enabling the user to run questions for each of these categories. Three indicative question-answer pairs, one for each question type, are in order: (Which was the birth place of Socrates?, Athens), (Is Nintendo located in Kyoto?, Yes!), (What is Parthenon?, the parthenon, a temple built in honor of athena...).

As we can see in Fig. 4, for the question "Which is the birth place of Lebron James?", LODQA returns the answer (i.e., "Ohio" in this example), and also a complete analysis for the question (see the right part of Fig. 4). Specifically, except for the short answer, one can find more information about its *provenance*, its *type* and its *confidence score*. Moreover, it returns the triple (in RDF format) where we found the question, whereas one can explore more information for each entity which is part of the triple. For example, as it is shown in the lower side of Fig. 4, one can explore for the entity "Lebron James" all its URIs (i.e., 13 URIs in total), the datasets where this entity occurs (i.e., 14 datasets) and one can have access to all the facts (i.e., 762 triples) about that person! Finally, a *tutorial video* is accessible in https://youtu.be/bSbKLlQBukk, whereas an *online demo* that will allow any user to ask questions will be released soon.

6   Conclusion

LODQA is a Question Answering approach that exploits hundreds of Linked Data sources, instead of using a single or few KBs (such as in [7, 22, 25, 32]). By using multiple datasets, the number of answerable questions increases, whereas the validity of any answer can be estimated from several sources, and answers are shown along with their provenance. We introduced an approach that exploits hundreds of datasets simultaneously and follows a variety of methods, without requiring any training data, for answering a question expressed in natural language. In particular, it includes methods for question cleaning, heuristic based question type identification, entity recognition, linking and disambiguation using Linked Data-based methods and pure NLP methods (specifically DBpedia Spotlight and Stanford CoreNLP), WordNet-based question expansion for tackling the lexical gap (between the input question and the underlying sources), and triple scoring for producing the final answer.

Concerning the evaluation, we used 1,000 questions of SimpleQuestions (v2) collection [6]. The evaluation show that regarding Entities Detection step, the combined method that exploits both entity recognition tools achieves the highest accuracy. This reflects the importance of using both KB-agnostic tools (SCNLP) and Large-scale KB-based tools (DBpedia Spotlight) for the tasks of entities recognition, linking and disambiguation. Regarding Answer Extraction, it seems that by using all the word expansion steps, we achieve the highest accuracy, while the method which does not perform any question words expansion (does not consider the lexical gap) achieves the worst results. This evidences the importance of tackling the lexical gap between the input question and the underlying sources for retrieving relevant information, i.e., triples. Moreover, we introduced experiments about the efficiency of our approach (e.g., half of the questions can be answered in less than 3.7 s) and the benefits of this approach in terms of answerable questions and answer verification (e.g., 28.4 million questions can be verified by at least two datasets). As a future work, we plan to improve the system for making it capable of returning responses to more complex questions, i.e., list questions or questions that require combining paths of triples.

Acknowledgements. The research work was supported by the Hellenic Foundation for Research and Innovation (HFRI) and the General Secretariat for Research and Technology (GSRT), under the HFRI PhD Fellowship grant (GA. No. 166).

References

1. Abujabal, A., Yahya, M., Riedewald, M., Weikum, G.: Automated template generation for question answering over knowledge graphs. In: Proceedings of the 26th International Conference on World Wide Web, pp. 1191–1200. International World Wide Web Conferences Steering Committee (2017)
2. Affolter, K., Stockinger, K., Bernstein, A.: A comparative survey of recent natural language interfaces for databases. arXiv preprint arXiv:1906.08990 (2019)
3. Bast, H., Haussmann, E.: More accurate question answering on freebase. In: Proceedings of the 24th ACM International on Conference on Information and Knowledge Management, pp. 1431–1440. ACM (2015)
4. Berant, J., Liang, P.: Imitation learning of agenda-based semantic parsers. Trans. Assoc. Comput. Linguist. **3**, 545–558 (2015)
5. Bollacker, K., Evans, C., Paritosh, P., Sturge, T., Taylor, J.: Freebase: a collaboratively created graph database for structuring human knowledge. In: Proceedings of the 2008 ACM SIGMOD International Conference on Management of Data, pp. 1247–1250. ACM (2008)
6. Bordes, A., Usunier, N., Chopra, S., Weston, J.: Large-scale simple question answering with memory networks. CoRR, abs/1506.02075 (2015)
7. Diefenbach, D., Singh, K., Maret, P.: WDAqua-core1: a question answering service for RDF knowledge bases. In: Companion of the The Web Conference 2018, pp. 1087–1091. International World Wide Web Conferences Steering Committee (2018)
8. Dimitrakis, E., Sgontzos, K., Tzitzikas, Y.: A survey on question answering systems over linked data and documents. J. Intell. Inf. Syst., 1–27 (2019). https://doi.org/10.1007/s10844-019-00584-7
9. Finkel, J.R., Grenager, T., Manning, C.: Incorporating non-local information into information extraction systems by Gibbs sampling. In: Proceedings of the 43rd Annual Meeting on Association for Computational Linguistics, pp. 363–370. Association for Computational Linguistics (2005)
10. Hakimov, S., Jebbara, S., Cimiano, P.: AMUSE: multilingual semantic parsing for question answering over linked data. In: d'Amato, C., et al. (eds.) ISWC 2017. LNCS, vol. 10587, pp. 329–346. Springer, Cham (2017). https://doi.org/10.1007/978-3-319-68288-4_20
11. Höffner, K., Walter, S., Marx, E., Usbeck, R., Lehmann, J., Ngonga Ngomo, A.-C.: Survey on challenges of question answering in the semantic web. Seman. Web **8**(6), 895–920 (2017)
12. Lehmann, J., et al.: DBpedia-a large-scale, multilingual knowledge base extracted from Wikipedia. Seman. Web **6**(2), 167–195 (2015)

13. Manning, C.D., Surdeanu, M., Bauer, J., Finkel, J.R., Bethard, S., McClosky, D.: The stanford coreNLP natural language processing toolkit. In: ACL (System Demonstrations), pp. 55–60 (2014)
14. Mendes, P.N., Jakob, M., García-Silva, A., Bizer, C.: DBpedia spotlight: shedding light on the web of documents. In: Proceedings of the 7th International Conference on Semantic Systems, pp. 1–8. ACM (2011)
15. Miller, G.A.: WordNet: a lexical database for English. Commun. ACM **38**(11), 39–41 (1995)
16. Mishra, A., Jain, S.K.: A survey on question answering systems with classification. J. King Saud Univ. Comput. Inf. Sci. **28**(3), 345–361 (2016)
17. Mountantonakis, M., Tzitzikas, Y.: High performance methods for linked open data connectivity analytics. Information **9**(6), 134 (2018)
18. Mountantonakis, M., Tzitzikas, Y.: LODsyndesis: global scale knowledge services. Heritage **1**(2), 335–348 (2018)
19. Mountantonakis, M., Tzitzikas, Y.: Large scale semantic integration of linked data: a survey. ACM Comput. Surv. (CSUR) **52**(5), 103 (2019)
20. Papangelis, A., Papadakos, P., Stylianou, Y., Tzitzikas, Y.: Spoken dialogue for information navigation. In: Proceedings of the 19th Annual SIGdial Meeting on Discourse and Dialogue, pp. 229–234 (2018)
21. Patra, B.: A survey of community question answering. CoRR, abs/1705.04009 (2017)
22. Radoev, N., Tremblay, M., Gagnon, M., Zouaq, A.: Answering natural language questions on RDF knowledge base in French. In: 7th Open Challenge in Question Answering over Linked Data (QALD 2017), Portoroz, Slovenia (2017)
23. Reddy, S., et al.: Transforming dependency structures to logical forms for semantic parsing. Trans. Assoc. Comput. Linguist. **4**, 127–140 (2016)
24. Rodrigo, A., Peñas, A.: A study about the future evaluation of question-answering systems. Knowl. Based Syst. **137**, 83–93 (2017)
25. Shekarpour, S., Marx, E., Ngomo, A.-C.N., Auer, S.: SINA: semantic interpretation of user queries for question answering on interlinked data. Web Seman. Sci. Serv. Agents World Wide Web **30**, 39–51 (2015)
26. Stockinger, K.: The rise of natural language interfaces to databases. In: ACM SIGMOD Blog (2019)
27. Tzitzikas, Y., Manolis, N., Papadakos, P.: Faceted exploration of RDF/S datasets: a survey. J. Intell. Inf. Syst. **48**, 1–36 (2016)
28. Wang, M.: A survey of answer extraction techniques in factoid question answering. In: Computational Linguistics, vol. 1, no. 1 (2006)
29. Yao, X., Berant, J., Van Durme, B.: Freebase QA: information extraction or semantic parsing. In: Proceedings of ACL (2014)
30. Yavuz, S., Gur, I., Su, Y., Srivatsa, M., Yan, X.: Improving semantic parsing via answer type inference. In: Proceedings of the 2016 Conference on Empirical Methods in Natural Language Processing, pp. 149–159 (2016)
31. Yih, W.-T., Chang, M.-W., He, X., Gao, J.: Semantic parsing via staged query graph generation: question answering with knowledge base. In: Proceedings of the Association for Computational Linguistics and the 7th International Joint Conference on Natural Language Processing, vol. 1, pp. 1321–1331 (2015)
32. Zhang, Y., He, S., Liu, K., Zhao, J.: A joint model for question answering over multiple knowledge bases. In: AAAI, pp. 3094–3100 (2016)

From Publications to Knowledge Graphs

Panos Constantopoulos[1,2(✉)] and Vayianos Pertsas[1,2]

[1] Department of Informatics, Athens University of Economics and Business, Athens, Greece
{panosc,vpertsas}@aueb.gr
[2] Digital Curation Unit, IMSI-Athena Research Centre, Marousi, Greece

Abstract. We address the task of compiling structured documentation of research processes in the form of knowledge graphs by automatically extracting information from publications and associating it with information from other sources. This challenge has not been previously addressed at the level described here. We have developed a process and a system that leverages existing information from DBpedia, retrieves articles from repositories, extracts and interrelates various kinds of named and non-named entities by exploiting article metadata, the structure of text as well as syntactic, lexical and semantic constraints, and populates a knowledge base in the form of RDF triples. An ontology designed to represent scholarly practices is driving the whole process. Rule -based and machine learning- based methods that account for the nature of scientific texts and a wide variety of writing styles have been developed for the task. Evaluation on datasets from three disciplines, Digital Humanities, Bioinformatics, and Medicine, shows very promising performance.

Keywords: Information extraction · Process mining · Knowledge base creation · Machine learning · Ontology population

1 Introduction

The explosion of publications in all disciplines makes it increasingly difficult for experts to maintain an overview of their domain and makes it harder to relate ideas from different domains [1, 2]. This situation could be significantly alleviated by supporting inquiries such as: find all papers that address a given problem; how was the problem solved; which methods are employed by whom in addressing particular tasks; etc. Answering such queries requires access to information about scholarly activities. Such information could be compiled interactively or automatically extracted from publications. Widely used search engines mostly leverage article metadata, while knowledge expressed in the actual text is only superficially exploited mostly by matching query terms to documents [3]. Understanding and encoding the knowledge contained in research articles is a complex task posing several challenges. The context of the research reported in an article needs to be expressed in a schema using information from the metadata of the article so that other occurrences in the same context can be associated to it. In addition, the text of the article is processed in order to extract concepts relevant to the documentation of research processes, which are subsequently associated according to predefined semantic relations.

© Springer Nature Switzerland AG 2020
G. Flouris et al. (Eds.): ISIP 2019, CCIS 1197, pp. 18–33, 2020.
https://doi.org/10.1007/978-3-030-44900-1_2

In this paper we present a process for creating knowledge graphs from publications. Information concerning scholarly activities or practices and their employed methods is extracted from the text of the publications. The extracted entities are associated with contextual information derived from publication metadata or other linked data repositories (e.g. ORCID or DBpedia) and everything is republished as a linked data knowledge graph capable of supporting complex queries of the form: *'who'* does *'what'*, *'why'* and *'how'*. Such knowledge graphs can promote digitally supported scholarly work, the integration of digital content, tools and methods, and enable the systematic codification and organization of work with respect not only to commonalities but also differences across disciplines, methodological traditions, and communities of researchers.

The conceptual model underlying the process is the domain-neutral *Scholarly Ontology (SO)* [4]. *SO* is specifically designed to capture different aspects of the scholarly process with core concepts like *'Activity'* denoting the research processes, *'Method'* the employed methods, *'Actor'* the activity's participants and *'Goal'* and *'Proposition'* the activity's objectives and results respectively. For the creation of SO-driven knowledge bases we developed *Research Spotlight (RS)*, a modular system that incorporates workflows for automatic retrieval of research articles from various APIs, distance supervision techniques for automatic annotation of training corpora, as well as modules for linked data generation/integration and entity/relation extraction [11]. For the latter we have implemented machine learning methods with specially designed feature spaces for extracting from text research activities, methods employed in those activities, and sequence relations between activities [22]. Several other relations specified in the ontology, such as the participation of actors in activities, the topic of a publication, the subject of a method, the goal of an activity, etc. are also captured either from the text using rule-based methods or by leveraging metadata and external sources.

So, *RS* generates linked, contextualized, structured data describing research activities and their outcomes, thus addressing the growing need for integrated access to information scattered in different publications. The performance of *RS* was evaluated with datasets from three different disciplines: Digital Humanities, Bioinformatics, and Medicine. We measured Precision, Recall and F1 scores in token- and entity-based evaluations with very promising results, indicating the potential for creating reliable research process knowledge bases. The results also confirmed the contribution of the specially designed features in achieving that performance.

The rest of this paper proceeds as follows: in Sect. 2 we present related work and explain how our task is different; in Sect. 3 we outline the ontology; in Sect. 4 we describe the knowledge base creation process; in Sect. 5 we present the methods used for extracting entities and relations from text; in Sect. 6 the evaluation experiments are briefly discussed; and we conclude in Sect. 7.

2 Related Work

To the best of our knowledge, the task of extracting research process information from scientific articles and republishing it as Linked Data, as introduced in [11, 22] and described in this paper, has not been previously addressed. That said, however, several past efforts aimed at extracting information from text based on an existing ontology. In [5] RDF triples are extracted from RSS feeds and published as Linked Open Data using mappings to DBpedia entities. The focus is on statistical methods and rules based on lexical form ignoring syntactic dependencies of tokens in the sentence that could allow better context understanding. The DBpedia project itself [6] is a significant operation to automatically extract knowledge from Wikipedia pages and info-boxes involving various NLP and feature-matching extractors that create RDF triples as instances of the DBpedia ontology. Here predefined rules are based on the DBpedia schema, metadata mappings, statistics of page links or word counts, and a number of feature extractors that exploit xml/html tags. However, the lexical, syntactic or structural analysis of raw text is not supported. In [8] a knowledge base is created with information extracted from French news wires by linking extracted entities to the instances of an ontology that unifies the models of GeoNames and Wikipedia and contains entities of type Person, Organization or Location retrieved from these sources. Common types of named entities are recognized and aligned with an existing database. In [9], a knowledge base is constructed by semi-automatic extraction of relations, based on the PRIMA ontology for risk management and a combination of machine learning techniques and predefined handcrafted rules. Syntactic dependencies that could yield patterns exploiting the deeper syntactic structure of sentences are not considered. Finally, in [10] an ontology-based information extractor employs handcrafted rules in order to extract soccer-related entities from various Web sources and map them onto soccer-specific semantic structures. The recognition of named entities is based solely on named entity lists, thus not supporting recognition of entities that are absent from the lists.

On the other hand, information extraction (IE) from scientific papers has attracted a lot of interest over the past years, as testified by the recent creation of a challenge on Scientific Information Extraction (ScienceIE) [3], the ACL RD-TEC Reference Dataset for Terminology Extraction and Classification [13], or domain-specific competitions such as BioCreAtIve[1]. Several recent works deal with extraction of key-phrases denoting tasks, scientific methods and materials from research documents [14, 15], association of extracted rhetorical entities, and named entities with linked data [16–18], or recognition of biomedical entities such as genes [19, 20]. They use features based on surface form, POS tags, or word embeddings and they employ classifiers such as SVMs, CRFs or neural networks, to extract key-phrases and named entities from text, as well as binary lexical semantic relations (synonym-of, hyponym-of). In [21], key-phrases denoting the "Focus", "Technique" and "Domain" of the articles are identified on the basis of syntactic patterns matched via rules to the dependency tree of each sentence in article abstracts. In [23] sentences from abstracts in the domains of clinical trials and biomedicine are classified in categories, such as introduction, purpose, method, results and conclusion, using various bag-of-words or bag-of-n-grams representations. A specialized system for extracting specific elements from legal contracts [7] uses sliding window classifiers and handcrafted features combined with word and POS tag embeddings to extract contract elements such as title, date, signatories' names, etc. In [26], authors and organizations are identified in scientific papers via CRFs using features that mainly deal with token surface form (lower/upper case, presence in gazetteers, font size, etc.) or structural characteristics (appearance in sections/paragraphs, first word in line, etc.). The extracted entities are

---

[1] http://biocreative.sourceforge.net/index.html.

then interrelated by further extracting the *hasAffiliation* property. For that, an SVM with Gaussian kernel is used with features related to the author affiliation markers and the distance of extracted strings.

In works related to action sequencing, such as [27], abstractions of action sentences are created based on a predefined template, which are then clustered together based on a functional similarity measure. In [28] deep reinforcement learning to extract sequences of labeled actions from sentences; each action is represented by verb-object predicates (e.g. cook(rice)) and the sequence relations are selected or eliminated based on their type (optional, exclusive or essential). In [29] a predefined list of names is used to map action descriptions and to interpret them as action sequences, or to generate navigational action descriptions using an encoder-aligner-decoder structure. Unlike the above methods, we associate actions that are not identified by single words or mapped to a fixed template or list of names. Instead, in our work actions have complex textual representations of variable length and cannot be labeled with words from a name-list. Moreover, we are not confined to deriving sequence relations from single lexical keywords. Instead, sequence relations are inferred from a combination of the actual textual context of activities along with structural properties of the text (e.g., relative positions of the entities in the texts).

In all of the approaches reviewed, IE from text uses either rules or ML methods based on features that handle mainly the surface form of words disregarding other information, such as attributes derived from syntactic dependencies or more complex syntactic patterns. Those ML methods perform inadequately in extracting research activities from text, as suggested by the evaluation of our baseline method that uses similar features. This behavior can be attributed to the following characteristics of the task at hand: Research activities are entities denoted only by textual descriptions without specific surface form and of arbitrary length, observed to exceed 50 tokens and possibly extending over multiple sentences, unlike the situation in common Named Entity Recognition (NER) problems. The textual chunks can contain stopwords, such as determiners, prepositions etc. Although in other NLP tasks (e.g. text classification) stopwords are often excluded because they just add noise, here they are deemed necessary because they are often important parts of the extracted entities. Sequence relations between activities cannot be detected solely from lexical indicators in the text. Other attributes of the activities, including their relative position in text, actual textual representation, etc., are also employed in order to improve classification.

In the above context, the contributions of our work are: (1) An end-to-end solution for understanding "who has done what, how, why and with what results" from the text of research articles. (2) A domain-independent procedure that automatically creates annotated corpora for training named entity recognizers, especially useful for entities of "non-common" type. (3) A system that leverages semantic information, surface form as well as deep syntactic and structural text analysis in order to extract information using both machine learning and rule-based modules. (4) The way we address the complexity of the particular task by including dependency embeddings, special syntactic sequences of words in addition to the order of appearance in text, as well as specialized features dealing with lexico-syntactic patterns, as opposed to just word surface form, currently employed in other works related to extracting knowledge from scientific literature. (5) Applicability across domains since no domain-specific lexica or training corpora are required. Our methods are demonstrated with test sets from three disciplines, capturing a variety of writing styles. (6) Higher performance compared to common NER or rule-based solutions especially considering the fact that the limited datasets we had available do not allow for more sophisticated ML approaches (such as deep learning methods).

3   Conceptual Framework: The Scholarly Ontology

The conceptual model underlying *RS* is based on the *Scholarly Ontology (SO)* [4], a domain-independent framework for modeling scholarly activities and practices. The rationale behind *SO* is to support answering questions of the form "*who does what, when, and how*" in and across scholarly domains, so the ontology is built around the central notion of *activity* and combines three perspectives: the *agency* perspective, concerning actors and their goals; the *procedure* perspective, concerning the intellectual framework and organization of work; and the *resource* perspective, concerning the material and immaterial objects consumed, used or produced in the course of activities. We here briefly review a subset of core *SO* concepts that constitute the *RS* schema guiding the extraction as well as the structuring of information (see Fig. 1).

Activity (e.g. an evaluation, a survey, an archeological excavation, a biological experiment, etc.) represents real events that have occurred in the form of intentional acts carried out by actors. Sequence of activities and composition from sub-activities are represented by the *follows* and *partOf* relations respectively. The instances of the Activity class are real processes with specific results, as opposed to those of the *Method* class, which are specifications, or procedures for carrying out activities to address specific goals. *Actor* instances are entities capable of performing intentional acts they can be accounted or referenced for. Actors can participate in activities, actively or passively, in one or more

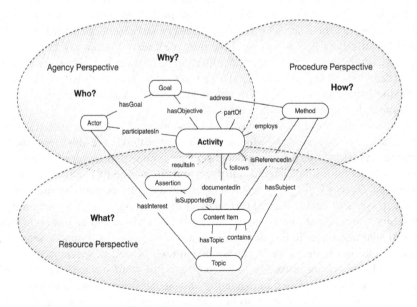

Fig. 1. The scholarly ontology core

roles. Subclasses of *Actor* are the classes *Person* and *Group* representing individual persons and collective entities respectively. Further specializations of *Group* are *Organization* and *Research Team*. *Content Item* comprises information resources, regardless of their physical carrier, in human readable form (e.g. images, tables, texts, mathematical expressions, etc.). *Assertion* includes all kinds of assertions in the scholarly domain and captures the intellectual essence of scholarly activity, comprising propositions resulting from activities; they can be *supportedBy* evidence provided by content items. Finally, *Topic* comprises thematic keywords expressing the subject of methods, the topic of content items, the research interests of actors, etc.

4 Knowledge Base Creation Process

An overview of the knowledge creation process is given in Fig. 2. The input consists of published -open access- research articles retrieved from repositories or Web pages in the preferred html/xml format. The format is exploited in extracting article metadata, such as authors' information, references and their mentions in text, legends of figures, tables etc. Entities, such as activities, methods, goals, propositions, etc., are extracted from the text of the article. These are associated in the relation extraction step, through various relations, e.g. *follows, hasPart, hasObjective, resultsIn, hasParticipant, hasTopic, has Affiliation,* etc. Encoded as RDF triples, these are published as linked data, using additional "meta-properties", such as *owl:sameAs, owl:equivalentProperty, rdfs:Label, skos:altLabel,* where appropriate.

Fig. 2. Knowledge base creation. Left to right: input, processes, extracted entities and relations

The entities targeted for extraction can be categorized into: (i) *named entities*, i.e. entities that have a proper name, such as instances of the SO classes *ContentItem, Person, Organization, Method* and *Topic*; and (ii) *nameless,* or *non-named entities*, identified by their own description but not given a proper name, such as instances of SO classes *Activity, Goal* and *Assertion*.

Different modules handle entities of each category. Figure 3 shows the architecture of RS implementing the above process. Input is obtained through:

(i)   SPARQL endpoints of various Web sources for creating Named Entities (NE) lists;
(ii)   user search keywords indicating the type of named entity to be recognized; and
(iii) URLs (e.g. journal Web pages that can be scraped) or publishers' APIs.

The main output of the system is the knowledge base published as linked data. The knowledge base creation process consists of two phases: (1) Preprocessing, for creating named entities lists and training the NER classifier and (2) Main Processing for the actual information extraction and publishing.

In *Preprocessing*, information is retrieved from sources such as DBpedia in order to build lists of named entities through the *NE List Creation* module. Specific queries using these entities are then submitted to the sources via the *API Querying* module. Retrieved articles are processed by the *Text Cleaning* module and the raw text at the output is added to a training corpus through the *Automatic Annotation* module that uses the entries of *NE list* to spot named entities in the text. The annotated texts are used to train a classifier to recognize the desired type of named entities.

Main Processing begins with harvesting research articles from Web sources, either using their APIs or by scraping publication Web sites. The articles are scanned for metadata which are mapped to SO instances according to a set of rules. In addition, specific html/xml tags inside the articles indicating images, tables and references are extracted and associated with appropriate entities according to SO, while the rest of the

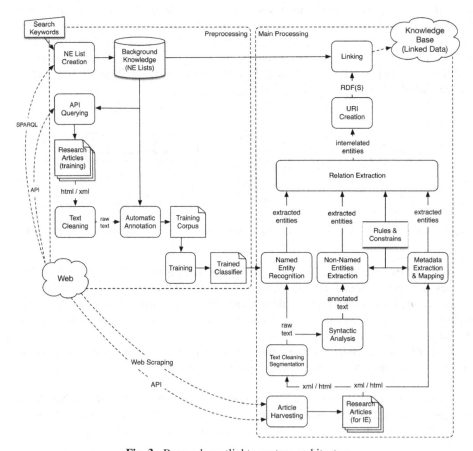

Fig. 3. Research spotlight - system architecture

unstructured, "raw" text is cleaned and segmented into sentences by the *Text Cleaning & Segmentation* module. The unstructured, "raw" text of the article is then input into the *Named Entity Recognition* module, where named entities of specific types are recognized. The segmented text is also inserted into a dependency parser using the *Syntactic Analysis* module. The output consists of annotated text in the form of dependency trees based on the internal syntax of each sentence, which is further processed by the *Non-Named Entities Extraction* module in order to extract text segments that contain other entities (such as Activities, Goals or Propositions). The output of the above steps (named entities, non-named entities and metadata) is fed into the *Relation Extraction* module that uses four kinds of rules: (i) syntactic patterns based on outputs of the dependency parser; (ii) surface form of words and POS tagging; (iii) semantic rules derived from SO; (iv) proximity constraints capturing structural idiosyncrasies of texts. Finally, based on the information extracted in the previous steps, URIs for the SO namespace are generated, and linked -when possible- to other strong URIs (such as the DBpedia entities stored in the named entities lists) in order to be published as linked data through a SPARQL endpoint.

For a detailed description of the steps of the above process see [11].

5 Extracting Entities and Relations

We initially developed a set of rule-based extraction algorithms that span the entire set of entities and relations of *SO* core. We subsequently applied machine learning methods starting with a critical subset of the *SO* core, namely activities and their sequence relations. To no surprise, the latter can outperform the former, but at the price of large training sets, hard to obtain. The outcomes of the rule-based methods, which require no training, can be used in building training sets and driving distance supervision methods, thus facilitating the bootstrapping of *RS* operation.

5.1 Rule-Based Extraction

Apart from "named entities" that can be identified using a NER (i.e. instances of *Method* class), we also need to extract "non-named" entities of highly variable length.

Textual chunks indicating *Activities*, *Goals* and *Assertions* are detected using syntactic analysis in conjunction with rules that exploit lexico-syntactic patterns derived from the reasoning frame of *SO* [4]. A dependency tree containing POS tags and syntactic dependencies for each word in a sentence is obtained using spaCy[2], a Python library offering industrial-strength Natural Language Processing functions. Each sentence is further analyzed using the semantic definitions of SO classes, the surface form of words, their POS tag and their syntactic dependencies.

---

[2] www.spacy.io.

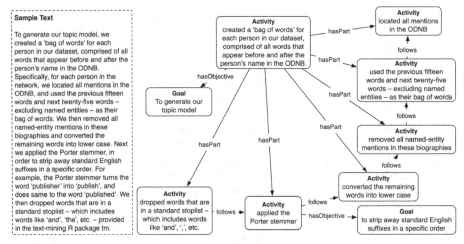

Fig. 4. Parthood and sequence relations

A sentence with verb in past or past/present perfect tense -in active or passive voice-containing no markers such as 'if' or 'that', quite likely describes an *Activity*, assuming the subject has the correct surface form ('we' or 'I' depending on the number of authors for active voice, no personal pronouns or determiners -to exclude vague subjects- for passive voice). Besides, 'that' following a verb can introduce a sub-sentence classified as *Assertion*, while a verb with dependent nodes with surface form 'to' or 'in order to' can introduce a sub-sentence classified as *Goal*.

The last step of information extraction involves detecting relations between previously extracted entities. *SO* semantics are employed for identifying the proper relation based on its domain and range. The organization of the text in sections and paragraphs induces proximity constraints enabling the inference of more complex, possibly inter-sentence, relations such as parthood and sequence of activities. The constraints used to identify relations are listed in Table 1. Relations marked with * are inherited from entity super-classes (Image, Table, Bib. Reference, Article from ContentItem; Person from Actor). The constraints for the *partOf* and *follows* relations (marked with **) can be relaxed in the presence of certain special indicators in the text (see Table 2). Parthood or sequence relations are assigned between the current and the last extracted activity either when a parthood or sequence indicator is detected, or by virtue of the relevant constraint. Figure 4 illustrates the extraction of sequence and parthood relations.

5.2 Extraction by Machine Learning Methods

We engineered several task-specific features exploiting the semantic context, syntactic dependencies of words and structural information, which we used in combination with word embeddings. The latter are dense vector representations of words that can be

Table 1. Types of constraints per relation type

Relation type	Semantic constrains (derived from SO)	Proximity Constraints – $P\_C()$ (from text structure)
isSupportedBy*	Domain: Proposition	XML/HTML pointers inside the Proposition chunk
	Range: Image|Table|Bibl. Reference	
partOf**	Domain: Activity	Co-occurrence with parent
	Range: Activity	Activity in the same paragraph
follows**	Domain: Activity	Co-occurrence with last Activity in the same paragraph
	Range: Activity	
contains*	Domain: Article	Co-occurrence in the same Article
	Range: Image|Table| Bibl. Reference	
participatesIn *	Domain: Person	Co-occurrence in the same Article
	Range: Activity	
employs	Domain: Activity	Co-occurrence in the same sentence
	Range: Method	
resultsIn	Domain: Activity	Co-occurrence in the same paragraph
	Range: Proposition	
hasObjective	Domain: Activity	Co-occurrence in the same sentence
	Range: Goal	
addresses	Domain: Method	Co-occurrence in the same sentence
	Range: Goal	
hasTopic*	Domain: Article	Co-occurrence in the same Article
	Range: Topic	
hasSubject	Domain: Method	Co-occurrence in the same Article
	Range: Topic	
hasInterest*	Domain: Person	Co-occurrence in the same Article
	Range: Topic	
hasGoal*	Domain: Person	Co-occurrence in the same Article
	Range: Goal	
isDocumentedIn*	Domain: Activity	Co-occurrence in the same Article
	Range: Article	
isReferencedIn*	Domain: Method	Co-occurrence in the same Article
	Range: Article	

produced in an unsupervised manner from unlabeled corpora and have proved instrumental in many NLP tasks. We actually employ three kinds of embeddings: word embeddings, part-of-speech (POS) tag embeddings, and dependency embeddings, all pre-trained for the domain of research processes, following the example of [7] where the first two kinds were combined.

We developed and compared several sliding window classifiers, thus exploring the activity and sequence extraction tasks along three dimensions:

(1) *Processing granularity*. We tested the effectiveness of classification at three levels of granularity: token-, sentence- and chunk-based classification.
(2) *Feature space*. The usual NLP practices were extended with the special features we developed and we assessed their effectiveness.
(3) *Machine learning method*. We developed classifiers employing Logistic Regression (LR) linear Support Vector Machines (SVM), and Random Forests (RF), as well as a 2-stage pipeline combination.

Table 2. Sequence and Parthood indicators along with their surface forms

Sequence and Parthood indicators	Surface forms
beginning_of_sequence	'first', 'initially', 'starting'
middle_of_sequence	'second', 'third', 'forth', 'fifth', 'sixth', 'then', 'afterwards', 'later', 'moreover', 'additionally', 'next'
end_of_sequence	'finally', 'concluding', 'lastly', 'last'
parthood_indicators	'specifically', 'first', 'concretely', 'individually', 'characteristically', 'explicitly', 'indicatively' 'analytically'

An unlabeled dataset obtained from 50,000 open-access research papers was used in order to create embeddings. The dataset consisted of approximately 10,000,000 sentences after metadata cleaning and parsing using spaCy, yielding 300,000,000 tokens and eventually a vocabulary of approx. 1,000,000 unique words. Word, part-of-speech tag (POS) and dependency (DEP) embeddings were generated from the above. Specifically: 100-dimensional word embeddings were produced using the Gensim implementation of word2vec[3] (skip-gram model); 25-dimensional POS embeddings were produced by replacing each token by its corresponding POS tag before running word2vec; and 25-dimensional DEP embeddings were produced by replacing each token by the label of the (unique) arc linking the token to its head in the dependency tree. Our experiments with other general-purpose, publicly available embeddings, such as those trained on the Common Crawl corpus using GloVe[4], or those trained on Wikipedia articles with word2vec, showed inferior performance compared to our domain-specific embeddings. This can be attributed to the fact that our embeddings are trained exclusively on scholarly articles, thus capturing the idiosyncrasies of scholarly writing styles.

For training the machine learning methods we used a labeled dataset derived from research articles randomly selected through publisher APIs (Springer and Elsevier), or by scraping online journals, e.g. Digital Humanities Quarterly. The training set, comprising

---

[3] https://radimrehurek.com/gensim/.
[4] https://nlp.stanford.edu/projects/glove/.

texts from 50 research articles covering 9 research domains, was annotated by two human annotators with an agreement of 81% kappa, yielding 1,700 activities comprising about 31,000 tokens, and 1,000 sequence relations. The intention was to create a training set spanning multiple disciplines in order to provide an adequate "baseline" training set for our methods. Our validation experiments proved this to be sufficient. On the other hand, any increase on either the size or the specialization of the dataset (e.g. developing the same ML method for a single discipline with larger training set) could arguably yield equal or better results.

Seven sliding window classifiers (SWC) and a 2-stage pipeline classifier were implemented for extracting research activities. They all perform token-based classification by examining each token t and its surrounding tokens in a fixed-size window, and classifying t as positive if it is part of a phrase expressing a research activity, or negative otherwise. The size of the window was set at 30 tokens around t (a total of 61 tokens) following tuning on the validation set. Zero-padding was used to represent tokens exceeding the sentence boundary. Each window of tokens is turned into a feature vector representing the token t being classified. In the above classifiers we tried combinations of three methods, Logistic Regression, linear Support Vector Machines and Random Forests, with different feature specifications. In addition to the above classifiers we implemented a two-stage pipeline. The first classifier, trained on all the sentences of the training set, detects only the existence of activities in the sentence without identifying their boundaries. The second classifier, trained only on sentences containing at least one activity, determines the boundaries of the chunks representing activities in the sentences classified as positive by the first classifier. The intuition is that, by splitting the task, each separate classifier will achieve high enough accuracy for their concatenation to produce better results, which was proven correct in the evaluation.

Extracting sequence relations requires examining all plausible activity pairs. For every pair of extracted activities the text chunk bounded by these two entities, [*act1*, ..., *act2*], empirically limited to 500 tokens, is considered to contain a candidate sequence relation. A binary classifier then determines whether the bounding activities of the chunk satisfy the property *follows*. We implemented three classifiers for extracting sequence relations between activities.

For details concerning the design of the classifiers in this section see [22].

6   Evaluation

The Information Extraction Modules of *RS* were evaluated by comparing their output with a "gold standard" produced by human annotators. According to established practice [24, 25], we generated the confusion matrices by comparing the output of the system with that of the human annotators and, using micro and macro-averaging, we calculated the precision, recall and F1 scores. We conducted two evaluation experiments: one "strict" and one "lenient", in which the confusion matrices were created based on "per-entity" and "per-token" calculations respectively.

Regarding **non-named entities** and their relations, our "gold standard" consisted of corpora produced from 50 articles annotated by two researchers. We drew from 29 different journals from various research areas (Digital Humanities, Geology, Medicine,

Bioinformatics, Biology, Computer Science, Sociology and Anthropology) to try our system with multiple writing styles. The non-named entities extracted belong to the classes *Activity*, *Goal* and *Proposition*, along with their relations *follows(act1, act2)*, *hasPart(act1, act2)*, *hasObjective(act, goal)*, *resultsIn(act, prop)*. Regarding **named entities** (instances of *Method* class) and the *employs(act, meth)* relation, the dataset was created in the pre-processing phase. For the experiments we used the Stanford NE[5] recognizer, trained/evaluated in the above dataset. The micro- and macro-averaged precision, recall and F1 scores based on confusion matrices from entity and token-based evaluation experiments, and individual scores for each type of entities and relations, are displayed in Tables 3, 4 and 5 respectively.

Regarding machine learning – based extraction, we evaluated the performance of all the classifiers by measuring Precision, Recall and F1 scores. After window-size selection and hyper-parameter tuning using 3-fold cross-validation on the training set, all the classifiers were trained on the entire training set. We used three different test sets from Digital Humanities, Bioinformatics and Medicine, presumably representing different writing styles, as well as their combination (ALL Test Set). Approximate Randomization Tests (ART) [30] between every classifier and the relevant baseline were carried out to ensure the statistical significance of the tests. Classifiers were grouped in zones of statistically similar results (shown by dividing lines in Tables 1 and 3) and ARTs were run on every combination of methods from different zones in order to ensure that the difference between any two measurements is statistically significant given our test sets. Results are shown in Tables 6 and 7. Confirming our intuition, the Pipeline classifier achieved the highest scores on every test set and criterion.

As a general comment, our more advanced features combined with domain specific embeddings significantly outperform other related methods (see Sect. 2) on the particular task at hand, as suggested by the evaluation of our baseline that simulates those by using similar features. For extensive accounts on the experimental evaluation of all the extraction methods see [11, 22].

Table 3. Micro & Macro Averaging Scores

	Macro-averaging			Macro-averaging		
	Precision	Recall	F1	Precision	Recall	F1
Entity-based	0.67	0.68	0.68	0.70	0.74	0.72
Token-based	0.87	0.77	0.81	0.84	0.83	0.83

---

[5] https://nlp.stanford.edu/software/CRF-NER.html.

Table 4. Entity Extraction

Entity Type	Entity-based			Token-based		
	P	R	F1	P	R	F1
Activity	0.70	0.75	0.72	0.79	0.85	0.81
Goal	0.74	0.78	0.76	0.86	0.74	0.80
Proposition	0.76	0.78	0.76	0.82	0.84	0.82
Method	0.80	0.69	0.74	0.75	0.72	0.73

Table 5. Relation Extraction

Relation type	P	R	F1
follows	0.69	0.72	0.71
hasPart	0.57	0.54	0.55
hasObjective	0.79	0.78	0.78
resultIn	0.54	0.58	0.56
employs	0.87	0.92	0.90

Table 6. Token-based Evaluation

		DH test set			BIOINF test set			MED test set			ALL test set		
		P	R	F1	P	R	F1	P	R	Fl	P	R	Fl
	Baseline	0.54	0.30	0.38	0.76	0.50	0.60	0.76	0.62	0.69	0.72	0.50	0.59
1	LR.WP.B	0.62	0.44	0.52	0.79	0.59	0.68	0.79	0.66	0.72	0.75	0.58	0.65
2	SVM.WP.B	0.60	0.50	0.54	0.80	0.66	0.72	0.78	0.68	0.73	0.74	0.63	0.68
3	LR.WPD.BS	0.78	0.76	0.77	0.83	0.81	0.82	0.88	0.83	0.85	0.84	0.80	0.82
4	SVM.WPD.BS	0.76	0.80	0.78	0.83	0.83	0.83	0.87	0.85	0.86	0.83	0.83	0.83
5	RF.PD.BS	0.79	0.80	0.80	0.85	0.83	0.84	0.89	0.83	0.86	0.85	0.82	0.83
6	LR.PD.S.BS	0.77	0.79	0.78	0.82	0.83	0.83	0.88	0.88	0.88	0.83	0.84	0.84
7	SVM.PD.S.BS	0.79	0.82	0.80	0.84	0.84	0.84	0.89	0.89	0.89	0.85	0.85	0.85
8	SVM-Pipeline	**0.83**	**0.82**	**0.82**	**0.87**	**0.89**	**0.88**	**0.90**	**0.93**	**0.92**	**0.87**	**0.89**	**0.88**

Table 7. Relation extraction evaluation

		DH test set			BIOINF test set			MED test set			ALL test set		
		P	K	Fl	P	R	Fl	P	R	Fl	P	R	Fl
	Baseline	0.62	0.72	0.67	0.65	0.89	0.76	0.59	0.92	0.72	0.62	0.88	0.72
1	LR(WPD)E-AVG-B	**0.87**	0.90	**0.88**	0.85	0.58	0.69	**0.94**	0.69	0.80	0.87	0.77	0.82
2	SVM(WPD)E-AVG-B	0.80	**0.93**	0.86	0.83	0.65	0.73	0.91	0.75	0.82	0.84	0.80	0.84
3	RF(PD)1H-SUM-B	0.81	**0.93**	0.87	**0.87**	**0.85**	**0.86**	**0.94**	**0.90**	**0.92**	**0.88**	**0.89**	**0.89**

7 Conclusion

We have presented a process, implemented in the *Research Spotlight* system, which leverages the *Scholarly Ontology* and deep syntactic analysis to extract information from articles and populate a knowledge base published as linked data. RS acquires information from the Web in several ways. Classifiers are automatically trained to recognize "non-common" named entities not supported by current serialized models. Using these together with the knowledge captured in the Scholarly Ontology, deep syntactic text analysis and machine learning methods, the system achieves extracting entities and relations representing research processes at a level of detail and complexity not addressed before.

Future work includes extracting further concepts for documenting research processes according to the Scholarly Ontology, such as goals, research questions, propositions, methods, etc., along with their corresponding relations (such as *partOf, employs, hasObjective*, etc.) and experimenting with more complex classifiers (e.g. CNNs or RNNs) when additional larger training datasets become available.

References

1. Bornmann, L., Mutz, R.: Growth rates of modern science: a bibliometric analysis based on the number of publications. J. Assoc. Inf. Sci. Technol. **66**, 2215–2222 (2015)
2. Renear, A.H., Palmer, C.L.: Strategic reading, ontologies, and the future of scientific publishing. Science **325**, 828–832 (2009)
3. Augenstein, I., Das, M., Riedel, S., Vikraman, L., McCallum, A.: SemEval 2017 task 10: ScienceIE, pp. 546–555 (2017)
4. Pertsas, V., Constantopoulos, P.: Scholarly ontology: modelling scholarly practices. Int. J. Digit. Libr. **18**, 173–190 (2017)
5. Gerber, D., Hellmann, S., Bühmann, L., Soru, T., Usbeck, R., Ngonga Ngomo, A.-C.: Real-time RDF extraction from unstructured data streams. In: Alani, H., et al. (eds.) ISWC 2013. LNCS, vol. 8218, pp. 135–150. Springer, Heidelberg (2013). https://doi.org/10.1007/978-3-642-41335-3_9
6. Lehmann, J., et al.: DBpedia - a large-scale, multilingual knowledge base extracted from Wikipedia. Semant. Web **6**, 167–195 (2015). https://doi.org/10.3233/SW-140134
7. Chalkidis, I., Michos, A., Androutsopoulos, I.: Extracting contract elements. In: ICAL, London, p. 10 (2017)
8. Stern, R., Sagot, B.: Population of a knowledge base for news metadata from unstructured text and web data. In: AKBC-WEKEX 2012, Montreal, Canada, pp. 35–40 (2012)
9. Makki, J., Alquier, A.-M., Prince, V.: Ontology population via NLP techniques in risk management. Int. J. Humanit. Soc. Sci. **3**, 212–217 (2008)
10. Buitelaar, P., Cimiano, P., Frank, A., Hartung, M., Racioppa, S.: Ontology-based information extraction and integration from heterogeneous data sources. Int. J. Hum. Comput. Stud. **66**, 759–788 (2008). https://doi.org/10.1016/j.ijhcs.2008.07.007
11. Pertsas, V., Constantopoulos, P.: Ontology-driven information extraction from research publications. In: Méndez, E., Crestani, F., Ribeiro, C., David, G., Lopes, J.C. (eds.) TPDL 2018. LNCS, vol. 11057, pp. 241–253. Springer, Cham (2018). https://doi.org/10.1007/978-3-030-00066-0_21
12. Goldberg, Y.: A primer on neural network models for natural language processing. J. Artif. Intell. Res. **57**, 345–420 (2015)
13. QasemiZadeh, B., Schumann, A.-K.: The ACL RD-TEC 2.0: a language resource for evaluating term extraction and entity recognition methods. In: Proceedings of the 10th Edition of the Language Resources and Evaluation Conference, pp. 1862–1868 (2016)
14. Lee, L.-H., Lee, K.-C., Tseng, Y.-H.: The NTNU system at SemEval-2017 task 10: extracting keyphrases and relations from scientific publications using multiple CRFs. In: 11th International Workshop on Semantic Evaluation (SemEval 2017), pp. 950–954 (2017)
15. Luan, Y., Ostendorf, M., Hajishirzi, H.: Scientific information extraction with semi-supervised neural tagging, pp. 2631–2641 (2017)

16. Sateli, B., Witte, R.: What's in this paper? Combining rhetorical entities with linked open data for semantic literature querying. In: Proceedings of the 24th International Conference on World Wide Web, pp. 1023–1028. ACM (2015)

17. Osborne, F., de Ribaupierre, H., Motta, E.: TechMiner: extracting technologies from academic publications. In: Blomqvist, E., Ciancarini, P., Poggi, F., Vitali, F. (eds.) EKAW 2016. LNCS (LNAI), vol. 10024, pp. 463–479. Springer, Cham (2016). https://doi.org/10.1007/978-3-319-49004-5_30

18. Sateli, B., Witte, R.: Semantic representation of scientific literature: bringing claims, contributions and named entities onto the Linked Open Data cloud. PeerJ Comput. Sci. **1**, e37 (2015)

19. Song, Y., Yi, E., Kim, E., Lee, G.G., Park, S.J.: POSBIOTM-NER: a machine learning approach for bio-named entity recognition, Korea, 305–350 (2004)

20. Plake, C., et al.: A support vector classifier for gene name recognition. In: BioCreAtIvE Workshop, Granada, Spain, pp. 1–5 (2004)

21. Gupta, S., Manning, C.: Analyzing the dynamics of research by extracting key aspects of scientific papers. In: Proceedings of 5th International Joint Conference on Natural Language Processing, pp. 1–9 (2011)

22. Pertsas, V., Constantopoulos, P., Androutsopoulos, I.: Ontology driven extraction of research processes. In: Vrandečić, D., et al. (eds.) ISWC 2018. LNCS, vol. 11136, pp. 162–178. Springer, Cham (2018). https://doi.org/10.1007/978-3-030-00671-6_10

23. Ruch, P., et al.: Using argumentation to extract key sentences from biomedical abstracts. Int. J. Med. Inf. **76**, 195–200 (2007)

24. Manning, C.D., Raghavan, P., Schutze, H.: Introduction to Information Retrieval. Cambridge University Press, New York (2008)

25. De Sitter, A., Calders, T., Daelemans, W.: A formal framework for evaluation of information extraction, University of Antwerp (2004)

26. Do, H.H.N., Chandrasekaran, M.K., Cho, P.S., Kan, M.-Y.M.Y.: Extracting and matching authors and affiliations in scholarly documents. In: Proceedings of the 13th ACM/IEEE-CS Joint Conference on Digital Libraries, JCDL 2013, p. 219 (2013)

27. Lindsay, A., Read, J., Ferreira, J.F., Hayton, T., Porteous, J., Gregory, P.: Framer: planning models from natural language action descriptions. In: Proceedings ICAPS, pp. 434–442 (2017)

28. Feng, W., Zhuo, H.H., Kambhampati, S.: Extracting action sequences from texts based on deep reinforcement learning (2018)

29. Mei, H., Bansal, M., Walter, M.R.: Listen, attend, and walk: neural mapping of navigational instructions to action sequences (2015)

30. Yeh, A.: More accurate tests for the statistical significance of result differences. In: Coling 2000 (2000)

Data Analytics

Analytics over RDF Graphs

Maria-Evangelia Papadaki[1,2]([✉]), Yannis Tzitzikas[1,2], and Nicolas Spyratos[3]

[1] Institute of Computer Science, FORTH, Heraklion, Greece
{marpap,tzitzik}@ics.forth.gr
[2] Computer Science Department, University of Crete, Heraklion, Greece
[3] Laboratoire de Recherche en Informatique, Université de Paris-Sud, Orsay, France
spyratos@lri.fr

Abstract. The continuous accumulation of multi-dimensional data and the development of Semantic Web and Linked Data published in RDF bring new requirements for data analytics tools. Such tools should take into account the special features of RDF graphs, exploit the semantics of RDF and support flexible aggregate queries. In this paper, we present an approach for applying analytics to RDF data, based on a high-level functional query language called HIFUN. According to that language, each analytical query is considered as a well-formed expression of a functional algebra and its definition is independent of the nature and structure of the data. In this work, we detail the required transformations, as well as the translation of HIFUN queries to SPARQL and we introduce the primary implementation of a tool, developed for these purposes.

Keywords: Analytics · RDF · Linked data

1 Introduction

The amount of data available on the Web today is increasing rapidly due to successful initiatives, such as the Linked Open Data movement[1]. More and more data sources are being exported or produced using the Resource Description Framework (or RDF, for short) standardized by the W3C[2]. SPARQL[3], which is the standard query language for RDF data, supports complex querying using regular path expressions, grouping, aggregation, etc., but the application of analytics to RDF data and especially to large RDF graphs is not so straightforward. The structure of such graphs tends to be complex, due to several factors: (a) different resources may have different sets of properties, (b) properties can be multi-valued (i.e. there can be triples where the subject and predicate are the same but the objects are different) and (c) resources may or may not have types. In addition, the analytical tools that have been developed, are not capable of supporting RDF graph analytics effectively, as they (i) focus on relational data,

---

[1] http://lod-cloud.net/.
[2] https://www.w3.org/RDF/.
[3] https://www.w3.org/TR/rdf-sparql-query/.

© Springer Nature Switzerland AG 2020
G. Flouris et al. (Eds.): ISIP 2019, CCIS 1197, pp. 37–52, 2020.
https://doi.org/10.1007/978-3-030-44900-1_3

(ii) can only work with a single homogeneous data set, (iii) neither support multiple central concepts, nor RDF semantics, (iv) demand deep knowledge of specific query languages, depending on data's structure and (v) do not offer flexible choices of dimension, measure, and aggregation.

In view of the above challenges, there is a need for a common formal framework that can be applied to one or more linked data sets and demands no programming skill. Motivated from this need, we are investigating an approach based on a high-level query language, called HIFUN [23], for applying analytics to RDF graphs. We study how that language can be applied to RDF data by clarifying how the concept of analysis context can be defined, what kind of transformations are required and how HIFUN queries can be translated to SPARQL. Moreover, we describe the primary implementation of an analytical tool based on the above.

The remainder of the paper is organized as follows: Sect. 2 describes the requirements and the related work. Section 3 introduces the related background knowledge. Section 4 focuses on how HIFUN can be applied to RDF data. Section 5 describes how HIFUN queries can be translated to SPARQL. Section 6 refers to application issues and describes the current implementation, and finally Sect. 7 concludes this work and discusses issues for future research.

2 Requirements and Related Work

2.1 Requirements

Today, there are domain-specific semantic warehouses, such as in the marine domain [26], or the cultural domain [10], as well as general-purpose knowledge bases, such as DBpedia and WikiData[4] (see [18] for a survey). These warehouses store huge volumes of integrated data from two or more disparate sources and their data can be analyzed in various ways. For example, the data warehouse of [26] contains data describing fish species according to several perspectives including water areas, countries, families, etc. Such data can be analyzed in order to find the number of different fish species by country or by water area. General-purpose knowledge bases on the other hand, such as DBpedia, can be analyzed for various purposes, (e.g. for finding the number of French actors, born in 1980). Apart from the above, analytics can be useful also for checking the quality of semantic integration activities, (e.g. for measuring the commonalities between several data sets as in [14,17]).

So, we need a way for applying analytics to any kind of RDF graph, - not only to multidimensional data expressed in RDF, but also to domain-specific or semantic data of general-purpose; a way, that would be applicable to several RDF data sets, as well as to any data source. We need an analytical tool that allows the user to select the desired data set(s) or desired parts thereof, formulate an analytic query without having any programming knowledge and finally get the results in the form of tables, plots or any other kind of visualization, intuitively.

---

[4] https://www.wikidata.org.

2.2   Related Work

There are many cases where it would be useful to publish multi-dimensional data (such as statistics) on the web in such a way that it can be linked to related data sets and concepts. The RDF Data Cube vocabulary[5] (QB) provides a means to publish such data on the web using the W3C RDF standard. That vocabulary consists of three main components: (i) the *measures*, which are the observed values of primary interest, (ii) the *dimensions*, which are the value keys that identify the measure and (iii) the *attributes*, which are the metadata. However, even though this vocabulary can be used for structuring and publishing multi-dimensional data, it cannot be used for applying analytics over it. In view of this limitation, several approaches have been proposed.

These approaches could be divided into two major groups: (i) those that extract Multi-dimensional Data (MD), that is data related to more than two dimensions from the web and load it into traditional data management systems for OLAP analysis [11,19], and (ii) those that perform OLAP analysis directly over the Semantic Web data, representing MD data in RDF [1].

The work in [12] analyzes data expressed in the RDF Data Cube format by constructing OLAP queries, which are then transformed into SPARQL. However, the proposed method requires a ROLAP engine to execute the OLAP queries and analyzes the resulting cubes through classical OLAP operations. On the other hand, the work in [9] presents a framework for analyzing LOD data. It differs from our work since the proposed method is not based on the usage of dedicated OLAP cube vocabularies (e.g. RDF Data Cube Vocabulary). Instead, it stores the RDF data in property tables (PTs) and transforms linked data to relational data so that it can be exploited using typical OLAP systems.

The representation of MD data in RDF can further be organized in two categories: (i) those that are based on specialized RDF vocabularies [5,6] and (ii) those that implicitly define a data cube over existing RDF graphs[6]. Our work follows the first approach since we apply analytics over data that has been expressed in the RDF Data Cube format (although the objective is to be applicable to any RDF data set).

The work in [29] defines OLAP operations on analysis cubes over graphs. However, its approach does not support heterogeneous graphs, and thus it cannot handle multi-valued attributes (e.g., a person being both "Greek" and "French"), nor semantics. Additionally, [3] presents a graph model for OLAP directly on RDF graphs and an extension of SPARQL for OLAP querying. However, according to [6], it cannot be guaranteed that the cubes on RDF graphs are multi-dimensional compliant.

The existing methods can also be classified into (i) those that require programming knowledge for analyzing the data and (ii) those that do not deal with lower-level technicalities. The work in [27] presents a system for analytics over (large) graphs. It achieves efficient query answering, by dividing the graph into

---

[5]   https://www.w3.org/TR/vocab-data-cube/.
[6]   https://team.inria.fr/oak/projects/warg/.

partitions. However, in contrast to our work, the user should have some programming knowledge, since it is necessary to write a few lines of code to submit the query. The work in [28] presents a method for applying statistical calculations on numerical linked data. It stores the data in arrays and performs the calculations on the arrays' values. Nevertheless, contrary to our work, it requires deep knowledge of SPARQL for formulating the queries.

In order to overcome one's difficulty in background programming knowledge, high-level languages have been developed for data analysis, too. However, there has not been much activity in introducing high-level languages, suitable for analytics on RDF data. While general-purpose languages, such as PIG Latin [20] and HiveQL [25] can be used, they are not tailored to address the peculiarities of the RDF data model. Even though, [7,8] present high-level query languages enabling OLAP querying of an extended format of data cubes [6], they are only applicable to data already represented and published using a corresponding vocabulary. As a consequence, they fall short in addressing a wide variety of analytical possibilities in non-statistical RDF data sources. In addition, [20] proposes a high-level language that supports semantics. However, it is targeted at processing structured relational data, limiting its use for semi-structured data such as RDF. Further, it provides only a finite set of primitives that is inadequate for the efficient expression of complex analytical queries.

Finally, a survey that is worth mentioning is [4], which introduces warehouse-style RDF analytics. There are similarities with our approach, since each analytical schema node corresponds to an RDF class, while each edge corresponds to an RDF property. Nonetheless, since the facts are encoded as unary patterns, they are limited to vertices instead of arbitrary subgraphs (e.g. paths).

In conclusion, in contrast to the aforementioned works, we focus on developing a user-friendly interface, where the user will be able to apply analytics to RDF data without dealing with lower-level technicalities. We envision a system, that will not be based on any specialized vocabulary and will be capable of analyzing one or several linked data sets.

3   Background

3.1   Resource Description Framework (RDF) and Linked Data

The Resource Description Framework (RDF) [2,16] is a graph-based data model for linked data interchanging on the web. It denotes resources through the use of Uniform Resource Identifiers (URIs), or anonymous resources (blank nodes) and constants (Literals). This framework uses triples, which are statements of the form subject-predicate-object (s, p, o), in order to relate a resource with other resources or constants.

Definition 1 (RDF Triple, RDF Data set, RDF Graph). A *triple* is considered to be any element of $T = (U \cup B) \times (U) \times (U \cup B \cup L)$, where U, B and L denote the sets of URIs, blank nodes and literals, respectively. An *RDF graph* (or *RDF data set*) is any finite subset of T.   □

For instance, if "schema" is the URI prefix "https://schema.org/", then schema:FinancialProduct is the URI of the class FinancialProduct. If "myStore" is the URI prefix "https://myStore.com", then myStore:product1 is the URI of a particular product. Consequently, the statement (myStore:product1, rdf:type, schema:FinancialProduct) is a triple, indicating that myStore:product1 is an instance of the class schema:FinancialProduct. In addition, (myStore:product1, schema:purchaseDate, "2019-05-09") is a triple, denoting that the product was purchased by its owner in "2019-05-09". The set of URIs could be classified in three different subsets, (i) entities (e.g. myStore:product1), (ii) properties (e.g. schema:purchaseDate) and (iii) classes (e.g. schema:FinancialProduct). An entity can be a subject or object in a triple, a property is always a predicate, while a class can be found in the object of a triple and corresponds to the type-/category, in which an entity belongs to.

We shall use the example of Fig. 1 as our running example, throughout the paper. The representation is in RDF and it shows a delivery invoice, that took place at "branch3" in "2019-05-09". The product that was delivered to that branch in the quantity of "400", was the "product4" of "Hermes" brand. The founder of that brand is "Manousos", who is both Greek and French.

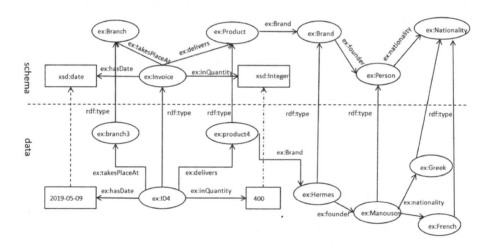

Fig. 1. Running example

3.2   HIFUN - A High Level Functional Query Language for Big Data Analytics

HIFUN [23] is a high-level functional query language for defining analytic queries over big data sets, independently of how these queries are evaluated. It can be applied over a data set that is structured or unstructured, homogeneous or heterogeneous, centrally stored or distributed.

Data Set Assumptions. To apply that language over a data set D, two assumptions should hold. The data set should (i) consist of *uniquely identified data items*, and (ii) have a set of *attributes*, each of which is viewed as a *function* associating each data item of D with a value, in some set of values. For example, if the data set D is a set of all delivery invoices over a year, in a distribution center (e.g. Walmart) which delivers products of various types in several branches, then the attribute "product type" (denoted as pt) is seen as a function $pt : D \rightarrow String$ such that, for each invoice i, $pt(i)$ is the type of product delivered according to the invoice i.

Definition 2 (Analysis Context). Let D be a data set and A be the set of all attributes $(a_1, ..., a_k)$ of D. An *analysis context* over D is any set of attributes from A, and D is considered the origin (or root) of that context. □

Note that an analysis context can consist of more than one roots. While one root means that data analysis concerns a single data set, the existence of two or more roots means that data analysis relates to two or more different data sets, possibly sharing one or more attributes.

A set of attributes could be represented as a directed labeled graph. Figure 2 shows our running example, expressed as such a context. From a syntactic point of view, the edges of it can be seen as triples of the form (source, label, target).

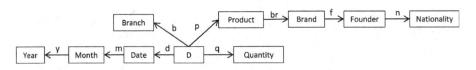

Fig. 2. Running example expressed as a HIFUN context

Direct & Derived Attributes. The attributes of a context are divided into two groups, the *direct* and the *derived*. The first group contains the attributes with origin D: these are the attributes whose values are given. The second group contains the attributes whose origins are different than D and whose values are computed based on the values of the direct attributes. For example, in Fig. 2 the attributes d, b, p and q are direct as their values appear on the delivery invoice, whereas m and y are derived, since their values can be computed from those of the attribute d (e.g. from the date 26/06/2019 one can derive the month 06 and the year 2019).

Definition 3 (HIFUN Analytic Query). A query in HIFUN is defined as an ordered triple $Q = (g, m, op)$ such that g and m are attributes of the data set D, having a common source (that is the *root D*) and op is an aggregate operation (or *reduction operation*) applicable on m-values. The first component of the triple is called *grouping function*, the second *measuring function* (or the *measure*) and the third *aggregate operation* (or *reduction operation*). □

The evaluation of such a query Q is done in a three-step process, as follows: (i) items with the same g-value g_i are grouped, (ii) in each group of items created, the m-value of each item in the group is extracted from D and (iii) the m-values obtained in each group are aggregated to obtain a single value v_i. Actually, the aggregate value v_i is the answer of Q on g_i. This means that a query is a triple of functions and its answer $AnsQ$ is a function, too.

4   Using HIFUN as an Interface to RDF Dataset

4.1   Motivation

There are several ways in which HIFUN can be used, such as for studying rewriting of analytic queries in the abstract [23] or for defining an approach to data exploration [24]. In this paper, we use HIFUN as a user-friendly interface for defining analytic queries over RDF data sets. To understand the proposed approach, consider a data source S with query language L (e.g. S could be a relational data set and L the SQL language). In order to use HIFUN as a user interface for S, we need to (a) define an analysis context, that is a subset D of S to be analyzed and some attributes of D that are relevant for the analysis and (b) define a mapping of HIFUN queries to queries in L.

Defining a subset D of S can be done using a query of L and defining D to be its answer (i.e. D is defined as a view of S); and similarly, the attributes that are relevant to the analysis can be defined based on attributes of D already present in S. However, defining a mapping of HIFUN queries to queries in L might be a tedious task. In [24] such mappings have been defined from HIFUN queries to SQL queries and from HIFUN queries to MapReduce jobs.

The main objective of this paper is to define a user-friendly interface allowing users to perform analysis of RDF data sets. To this end, we use the HIFUN language as the interface. In other words, we consider the case, where the data set S mentioned above is a set of RDF triples and its language L is the SPARQL language. Our main contributions are: (a) the proposal of tools for defining a HIFUN context from the RDF data set S and (b) defining a mapping from HIFUN queries to SPARQL queries. With these tools at hand, a user of the HIFUN interface can define an analysis context of interest over S and issue analytic queries using the HIFUN language. Each such query is then translated by the interface to a SPARQL query, which in turn is evaluated over the RDF triples of D and the answer is returned to the user.

4.2   Applicability of HIFUN

Recall that two assumptions must be satisfied in order to apply HIFUN (see Sect. 3.2). The first assumption is satisfied by the RDF data since each resource in RDF is identified by a distinct URI. Therefore, the data set D in HIFUN can be any subset of the set of all the available URIs. The second assumption, the functionality of attributes, is satisfied by the RDF properties, which are defined as functional (i.e. owl:FunctionalProperty) or are effectively functional (i.e. even if they are not declared as functional, they are single-valued for the resources in the data set D). Consequently, the cases that require special handling, include the following: (a) properties with no value, since HIFUN assumes that there are no empty values in the data, in contrast to RDF, where properties with no value may exist, (b) properties, that are multi-valued, and (c) definition of analysis contexts that correspond to a transformation of the original data.

These issues are discussed in the following sections.

4.3 How to Specify the Context of Analysis

In order to specify an analysis context, the classes and the properties of interest of an RDF graph should be selected. The user can select as the *root* of the analysis context any class of the RDF graph and as attribute any property of it (whose domain is that class). For example, any of the classes "ex:Invoice", "ex:Branch", "ex:Product", "ex:Brand", "ex:Person", "ex:Nationality" of Fig. 1 can be selected as the root of a context, while any of the properties "ex:hasDate", "ex:takesPlaceAt", "ex:delivers", "ex:inQuantity", "ex:brand", "ex:founder", "ex:nationality" as its attributes, since they do have a common root. An analysis context can also be defined by transforming the original data, and such issues are discussed in Sect. 6.

5 Translation of HIFUN Queries to SPARQL

In this section, we show how a HIFUN query can be translated to a SPARQL query. Recall, that a query in HIFUN is defined as an ordered triple $Q = (g, m, op)$, where g is the grouping function, m the measuring function and op the aggregate operation. On the other hand, an aggregate query in SPARQL is defined as:

```
SELECT ?group (function(?var) AS ?result)
WHERE {
 . . . .
}
GROUP BY ?group
```

Based on the above definitions, a HIFUN query Q can be encoded as a SPARQL group-by query, as shown in row 1 of Table 1. The grouping function corresponds to the projections in the SELECT clause as well as to the aggregate variable(s) in the GROUP BY clause, the reduction (or aggregate operation) to the aggregate SPARQL function, and the measuring function to the argument of that function. Note that, we define as target the *codomain* (or *target set*) of the HIFUN attributes. The answer to this query is a binary table with two variables, $target(g)$ and Res (Res is a user-defined variable that holds the aggregate result).

Definition 4 (Translation of a HIFUN Query to a SPARQL query).
A HIFUN query Q over a context C is translated to SPARQL by grouping
the items with the same projection value g_i, extracting the aggregate function's
argument value m_i of each item of the created groups and aggregating the m_i
values obtained in each group, in order to obtain a single value v_i. $\square$

Table 1. HIFUN to SPARQL

id	name	HIFUN query Q	SPARQL query
1	Plain	$(grouping,$ $measuring,$ $reduction)$	SELECT ?target(grouping) reduction(?target(measuring)) As ?Res WHERE { .... } GROUP BY ?target(grouping)
2	Plain	(b, q, SUM)	SELECT ?branch SUM(?quantity) AS ?TOTALS WHERE { ?ID ex:takesPlaceAt ?branch . ?ID ex:inQuantity ?quantity . } GROUP BY ?branch
3	Pairing	$(b \wedge p, q, SUM)$	SELECT ?branch ?product SUM(?quantity) AS ?TOTALS WHERE { ?ID ex:takesPlaceAt ?branch . ?ID ex:delivers ?product . ?ID ex:inQuantity ?quantity . } GROUP BY ?branch ?product
4	Complex grouping function	$((dom) \wedge b, q, SUM)$	SELECT ?branch ?mon SUM(?quantity) AS ?TOTALS WHERE { ?ID ex:hasDate ?date . ?date rdfs:label ?label . ?ID ex:inQuantity ?quantity . ?ID ex:delivers ?product . ?product ex:hasBrand ?branch . } GROUP BY ?branch (month(?label) AS ?mon)
5	Attribute-restricted	$(e/E, e', op)$	SELECT ?target(e), op(?target(e')) As ?Res WHERE { FILTER(op(?target(e')) op') } GROUP BY ?target(e)
6	Attribute-restricted	$(b/\text{"branch1"}, q, SUM)$	SELECT ?product SUM(?quantity) AS ?TOTALS WHERE { ?ID ex:takesPlaceAt ?branch. ?ID ex:delivers ?product. ?ID ex:inQuantity ?quantity. FILTER regex((?branch), "branch1", "i") } GROUP BY ?product
7	Result-restricted	$(e, e', op)/F$	SELECT ?target(e), op(?target(e')) As ?Res WHERE { ... } GROUP BY ?target(e) HAVING(op(?target(e')) op')
8	Result-restricted	$(b, q, SUM)/F,$ where $F = \{b_i \in Branch/ans(b_i) > 300\}$	SELECT ?branch SUM(?quantity) AS ?TOTALS WHERE { ?ID ex:takesPlaceAt ?branch. ?ID ex:inQuantity ?quantity. } GROUP BY ?branch HAVING (SUM(?quantity) > 300)

Example 1. Suppose, that we would like to find the total quantities of products, delivered to each branch during the year. This query would be expressed in HIFUN and SPARQL respectively, as it is shown in row 2 of Table 1.

Complex Grouping/Measuring Function. A grouping (as well as a measuring) function in HIFUN can be *more complex* using the following four operations on functions, that are defined in [23]: *pairing (∧)*, *composition (○)*, *Cartesian product projection (×)* and *restriction (//)*.

These operations form the so called functional algebra [22] and they are well known, elementary operations except probably for pairing, which works as a tuple constructor and is defined as follows:

Pairing: Let $f : X \rightarrow Y$ and $g : X \rightarrow Z$ be two functions with common domain X. The pairing of f and g, denoted $f \wedge g$ is a function from X to $Y \times Z$ defined by: $f \wedge g(x) = (f(x), g(x))$, for all x in X [23].

Example 2. Suppose, that we ask for the total quantities delivered by branch and product. The answer to this query Q is a function, associating each pair $(branch, product)$ with a total quantity. In other words, Q asks for the total quantities delivered by branch and product and it would be formulated as it is shown in row 3 of Table 1.

Example 3. Suppose we want the total quantities of products delivered, grouped by branch and month. Since the attribute of *month* is a derived one, it would have to be expressed using the operator of *composition*. That operator would be used to combine the attribute of *date* with that of *month*. On the other hand, the corresponding query in SPARQL would be expressed by "extracting" the value of month from the date by applying the built-in SPARQL function of *month*, which operates on date values. These queries would be defined as it is shown in row 4 of Table 1.

Restricted Query. A query Q in HIFUN can further be enriched by introducing functional restrictions either at the level of attributes or at the level of query answers.

Regarding the case of attribute-restricted queries, the restrictions are applied at the level of the attributes, filtering the results internally and the queries are defined as shown in row 5 of Table 1. The HIFUN query is evaluated by computing the restriction e/E (where E is any subset of the Data set D) and then, the query $(e/E, e', op)$ over E. On the other hand, in SPARQL the query is evaluated by specifying the subset of the data set D the user is interested in (using the $FILTER$ operator, triples patterns, etc.)

Example 4. Suppose, that we would like to find the total quantities of products, that received by a specific branch e.g. "branch1", group by product. Then, the queries would be expressed as shown in row 6 of Table 1. Alternatively, the restriction could be performed using a triple pattern, by specifying the particular branch that a delivery invoice took place. In that case, each branch would have been represented with a URI, e.g. $ex : branch_i$ and the constraint would be defined using triples of the form $?ex : ID\ ex : takesPlaceAt\ ex : branch_i$.

The decision between a triple pattern and filter for expressing a restriction in the inner results of a SPARQL query depends on the way our data has been represented. The first case concerns data that has been represented with $URIs$ and the use of triple patterns is preferred. The second one relates to data that has been produced using *literals* and in that case, the operator of *filter* is applied. This operator is also used in boolean conditions where any unwanted results should have to be filtered out.

Regarding the case of result-restricted queries, the final result can be filtered by setting restrictions on them. The queries, in this case, are defined as it is shown in row 7 of Table 1. The query in HIFUN is evaluated by first evaluating the query $Q = (e, e', op)$ over D and then computing the restriction $ansQ/F$. The corresponding SPARQL is evaluated and its result is filtered using the $HAVING$ operator.

Example 5. Suppose that we would like to find the number of products received by branch, but only for those branches that received more than 300 products. The corresponding queries would be expressed as shown in row 8 of Table 1.

6   Application and Implementation

6.1   Defining an Analysis Context over RDF Data

As discussed in Sect. 4.1, in order to define an analysis context D, the query language of the data source can be exploited. This is required mainly in general-purpose knowledge bases (as discussed in Sect. 2). Below, we describe some methods for defining such a context over RDF data. Here, we consider an analysis context as a pair (E, F), where E is a set of resources (i.e. a set of URIs), and F is a set of attributes for the objects in E. Such a pair can be defined, as follows:

- M_{plain}: If the data set D consists only of a single class (say C) and all the properties have as domain or range that class, the analyst has just to select the desired subset of these properties. In this case, $E = \{u \in U \mid (u, \text{rdf:type}, C)\}$, and F is any non empty subset of $Props(C) = \{u \mid (p,\text{rdfs:domain},C)\} \cup \{u \mid (p,\text{rdfs:range},C)\}$. Note that, this case captures data expressed in the RDF Data Cube format.
- M_{QLview}: The analyst can use a SPARQL SELECT query for defining a view, having all the attributes required for the analysis. For instance, if $v_1, \ldots v_k$ is the set of variables in the SELECT part of the query, then E can be considered to be the set of bindings of v_1, while F the bindings of the set of variables $v_2, \ldots, v_k$.
- $M_{QLobjects}$: The analyst can write a SPARQL query for defining only the objects of interest, i.e. the set E, not their attributes. Then, a tool could be used to suggest (or let the analyst select) the applicable properties F based on the schema and/or data. Also, note that the objects of interest E can be defined explicitly, i.e. by just providing the list of the desired URIs.

Special Cases. Regarding the $M_{QLobjects}$ case, note that if all the properties in F have a value for each E and they are single-valued too, then the context has already been defined. However, there are cases that may require special handling: (i) there are properties with no value (such cases can occur in M_{QLview} if the OPTIONAL keyword is used), (ii) the analyst is interested in a path of properties, not a single property, and (iii) there are properties, which are multi-valued. To tackle such cases, some transformations may be required. A few feature operators that could be used in such cases, are indicated in Table 2. That table lists the nine most frequent *Linked Data-based Feature Creation Operators* (for short *FCOs*), as defined in [15] and they have been re-grouped according to our requirements. T denotes a set of triples, P a set of properties and p, p_1, p_2 denote properties. In detail,

- fco_1 suits to the normal case, i.e. to properties that are functional, e.g. the date that each product was delivered, the branch where each invoice took place, and its value can be numerical or categorical.
- fco_2 and fco_3 are related to issues that concern missing and multi-valued properties.
- fco_4 can be used for transforming a multi-valued property to a set of single-valued features, e.g. one boolean feature for each nationality, that a founder may have.
- fco_5 and $fco6$ relate to the degree of an entity.
- fco_7 to $fco9$ investigate paths in an RDF graph, e.g. whether at least one founder of a brand is "French".

Consequently, the aforementioned cases could be handled by transforming our data set properly, using the feature operators already described. Specifically, regarding case (i), i.e. properties with empty values, the transformations 2 and 3 of Table 2 could be applied for turning such properties into integers. Concerning case (ii), the transformations 7 to 9 could be used for specifying a path (a sequence of properties $p_1, p_2, ..., p_n$ etc.) and handle it as an individual property p. Finally, relating the case (iii), the transformations 2 to 4 could be used for inspecting the existence of any multi-valued properties and the conversion of them to single-valued.

Application of HIFUN in Special Cases. Some of the denoted special cases can be handled in HIFUN without having previously transformed our data.

For example, regarding the case (ii), where the user may be interested in a path P of an RDF graph, the operator of the composition ($\circ$) can be used for expressing such a path in HIFUN i.e. by combining all the attributes of P.

Example 6. Suppose that we would like to find the quantity of sold products per month, that belong to brands of French founders. The query in HIFUN would be defined as, $((m \circ d)/E, q, SUM)$, where $E = \{x | x \in D \land (n \circ f \circ br \circ p)(x)) = French\}$.

Example 7. Suppose now that, we would like to find the sum of total sales of the month "September" grouped by the nationality of the founders. Such a

Table 2. Feature creation operators

id	Operator defining f_i	Type	$f_i(e)$
	Plain selection of one property		
1	p.value	num/categ	$f_i(e) = \{\, v \mid (e,p,v) \in \mathcal{T}\,\}$
	For missing values and multi-valued properties		
2	p.exists	boolean	$f_i(e) = 1$ if (e,p,o) or $(o,p,e) \in \mathcal{T}$, otherwise $f_i(e) = 0$
3	p.count	int	$f_i(e) = \lvert\{\, v \mid (e,p,v) \in \mathcal{T}\,\}\rvert$
	For multi-valued properties		
4	p.values.AsFeatures	boolean	for each $v \in \{\, v \mid (e,p,v) \in \mathcal{T}\,\}$ we get the feature $f_{iv}(e) = 1$ if (e,p,v) or $(v,p,e) \in \mathcal{T}$, otherwise $f_{iv}(e) = 0$
	General ones		
5	degree	double	$f_i(e) = \lvert\{(s,p,o) \in \mathcal{T} \mid s = e \text{ or } o = e\}\rvert$
6	average degree	double	$f_i(e) = \frac{\lvert triples(C)\rvert}{\lvert C\rvert}$ s.t. $C = \{\, c \mid (e,p,c) \in \mathcal{T}\,\}$ and $triples(C) = \{(s,p,o) \in \mathcal{T} \mid s \in C \text{ or } o \in C\}$
	Indicative extensions for paths		
7	p1.p2.exists	boolean	$f_i(e) = 1$ if $\exists\, o2$ s.t. $\{(e,p1,o1),(o1,p2,o2)\} \subseteq \mathcal{T}$
8	p1.p2.count	int	$f_i(e) = \lvert\{\, o2 \mid (e,p1,o1),(o1,p2,o2) \in \mathcal{T}\,\}\rvert$
9	p1.p2.value.maxFreq	num/categ	$f_i(e) =$ most frequent $o2$ in $\{\, o2 \mid (e,p1,o1), (o1,p2,o2) \in \mathcal{T}\,\}$

query would be expressed in HIFUN as, $((n \circ f \circ br \circ p)/E, q, SUM)$ where $E = \{x \mid x \in D \wedge (m \circ d)(x)) = September\}$.

As regards case (iii), we could apply HIFUN to multi-valued attributes, if we used "$\in$" instead of "$=$" in the restriction. Note that, a multi-valued attribute should always correspond to a terminal node of a HIFUN context.

Example 8. Suppose that, the attribute of "nationality" is a multi-valued property i.e. a person can have more than one nationalities, as shown in the example of Fig. 1 and we would like to find the quantity of sold products per month, that belong to brands of French founders. Then, this query would be defined in HIFUN as,

$$((m \circ d)/E, q, SUM), \text{ where } E = \{x \mid x \in D \wedge (n \circ f \circ br \circ p)(x)) \in French\}.$$

6.2   Implementation Issues

As regards the specification of the analysis context, the cases of M_{plain} and M_{QLview} (as well as the case of CSV files) can be supported by tools like Facetize [13], that also offer cleaning functionality, as well as the ability to organize the values of some dimensions, hierarchically. The case of $M_{QLobjects}$ can be supported by tools like LODSyndesisML [15], that reads a list of URIs and enrich

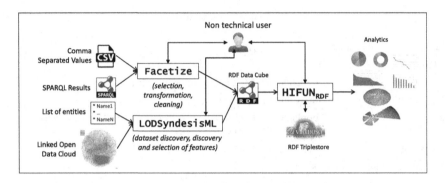

Fig. 3. A few indicative workflows involving HIFUN$_{RDF}$.

them with attributes, by exploiting several data sets published as Linked Data. The output of the above tools can be straightforwardly converted to the RDF Data Cube format.

In our work, we have currently developed a tool, called HIFUN$_{RDF}$ which applies the HIFUN query language to RDF data. For the time being, only data in the RDF Data Cube format is supported. In order to execute a query, the user should define: (1) the grouping function, (2) the measuring function, (3) the aggregate operation, and optionally, (4) set restrictions to the grouping, the measuring functions or to the final results. The inserted values are used to construct the corresponding HIFUN query, which is subsequently converted to the respective SPARQL. The latter is executed on the triple store OpenLink Virtuoso[7] (where the input data has been uploaded) and the returned results are saved in a .csv file, in the form of - $var_1, var_2, ..., var_i$, TOTALS. Indicative flows between HIFUN$_{RDF}$ and other tools are illustrated in Fig. 3.

Now, we are in the process of extending our application, to support analytics over RDF data in general (and not only to data expressed in the RDF Data Cube format). We are designing an interface, that will let the user specify the analysis context (as well as the required transformations) and formulate the HIFUN query, interactively. After that step, we will focus on the visualization of the results, probably by extending the visualization method presented in [21][8].

7   Concluding Remarks

In this paper, we examined the basics for applying HIFUN over RDF data. We described methods for defining the context of analysis over an RDF graph and we showed how HIFUN queries can be translated to SPARQL. Moreover, we described, in brief, a first implementation of the approach that, for the time being, can be applied to data expressed in the RDF Data Cube format. In the future, we plan to investigate, more complex queries and OLAP-style operations

---

[7] https://virtuoso.openlinksw.com/.

[8] http://www.ics.forth.gr/isl/3DLod/.

over RDF graphs including the interplay with hierarchies and inference. Besides, we plan to design a graphical user interface appropriate for applying HIFUN to RDF data, and to further work on the visualization part of the analytical results.

References

1. Abelló, A., et al.: Fusion cubes: towards self-service business intelligence. Int. J. Data Warehous. Min. (IJDWM) **9**, 66–88 (2013)
2. Antoniou, G., Van Harmelen, F.: A Semantic Web Primer. MIT Press, Cambridge (2004)
3. Beheshti, S.-M.-R., Benatallah, B., Motahari-Nezhad, H.R.: Scalable graph-based OLAP analytics over process execution data. Distrib. Parallel Databases **34**(3), 379–423 (2014). https://doi.org/10.1007/s10619-014-7171-9
4. Colazzo, D., Goasdoué, F., Manolescu, I., Roatiş, A.: RDF analytics: lenses over semantic graphs. In: Proceedings of the 23rd International Conference on World Wide Web (2014)
5. Etcheverry, L., Vaisman, A.A.: Enhancing OLAP analysis with web cubes. In: Simperl, E., Cimiano, P., Polleres, A., Corcho, O., Presutti, V. (eds.) ESWC 2012. LNCS, vol. 7295, pp. 469–483. Springer, Heidelberg (2012). https://doi.org/10.1007/978-3-642-30284-8_38
6. Etcheverry, L., Vaisman, A.A.: QB4OLAP: a new vocabulary for OLAP cubes on the semantic web. In: Proceedings of the Third International Conference on Consuming Linked Data (2012)
7. Etcheverry, L., Vaisman, A.A.: Querying semantic web data cubes. In: AMW (2016)
8. Etcheverry, L., Vaisman, A.A.: Efficient analytical queries on semantic web data cubes. J. Data Semant. **6**(4), 199–219 (2017). https://doi.org/10.1007/s13740-017-0082-y
9. Inoue, H., Amagasa, T., Kitagawa, H.: An ETL framework for online analytical processing of linked open data. In: Wang, J., Xiong, H., Ishikawa, Y., Xu, J., Zhou, J. (eds.) WAIM 2013. LNCS, vol. 7923, pp. 111–117. Springer, Heidelberg (2013). https://doi.org/10.1007/978-3-642-38562-9_12
10. Isaac, A., Haslhofer, B.: Europeana linked open data-data. europeana. eu. Semant. Web **4**, 291–297 (2013)
11. Kämpgen, B., Harth, A.: Transforming statistical linked data for use in OLAP systems. In: Proceedings of the 7th International Conference on Semantic Systems (2011)
12. Kämpgen, B., O'Riain, S., Harth, A.: Interacting with statistical linked data via OLAP operations. In: Simperl, E., et al. (eds.) ESWC 2012. LNCS, vol. 7540, pp. 87–101. Springer, Heidelberg (2015). https://doi.org/10.1007/978-3-662-46641-4_7
13. Kokolaki, A., Tzitzikas, Y.: Facetize: an interactive tool for cleaning and transforming datasets for facilitating exploratory search. arXiv preprint arXiv:1812.10734 (2018)
14. Mountantonakis, M., Tzitzikas, Y.: On measuring the lattice of commonalities among several linked datasets. Proc. VLDB Endow. **9**, 1101–1112 (2016)
15. Mountantonakis, M., Tzitzikas, Y.: How linked data can aid machine learning-based tasks. In: Kamps, J., Tsakonas, G., Manolopoulos, Y., Iliadis, L., Karydis, I. (eds.) TPDL 2017. LNCS, vol. 10450, pp. 155–168. Springer, Cham (2017). https://doi.org/10.1007/978-3-319-67008-9_13

16. Mountantonakis, M., Tzitzikas, Y.: LODsyndesis: global scale knowledge services. Heritage **1**, 335–348 (2018)
17. Mountantonakis, M., Tzitzikas, Y.: Scalable methods for measuring the connectivity and quality of large numbers of linked datasets. J. Data Inf. Qual. (JDIQ) **9**, 1–49 (2018)
18. Mountantonakis, M., Tzitzikas, Y.: Large scale semantic integration of linked data: a survey. ACM Comput. Surv. (CSUR) **52**, 1–40 (2019)
19. Nebot, V., Berlanga, R.: Building data warehouses with semantic web data. Decis. Support Syst. **52**, 853–868 (2012)
20. Olston, C., Reed, B., Srivastava, U., Kumar, R., Tomkins, A.: Pig Latin: a not-so-foreign language for data processing. In: Proceedings of the 2008 ACM SIGMOD International Conference on Management of Data (2008)
21. Papadaki, M.-E., Papadakos, P., Mountantonakis, M., Tzitzikas, Y.: An interactive 3D visualization for the LOD cloud. In: EDBT/ICDT Workshops (2018)
22. Spyratos, N.: A functional model for data analysis. In: Larsen, H.L., Pasi, G., Ortiz-Arroyo, D., Andreasen, T., Christiansen, H. (eds.) FQAS 2006. LNCS (LNAI), vol. 4027, pp. 51–64. Springer, Heidelberg (2006). https://doi.org/10.1007/11766254_5
23. Spyratos, N., Sugibuchi, T.: HIFUN - a high level functional query language for big data analytics. J. Intell. Inf. Syst. **51**(3), 529–555 (2018). https://doi.org/10.1007/s10844-018-0495-6
24. Spyratos, N., Sugibuchi, T.: Data exploration in the HIFUN language. In: Cuzzocrea, A., Greco, S., Larsen, H.L., Saccà, D., Andreasen, T., Christiansen, H. (eds.) FQAS 2019. LNCS (LNAI), vol. 11529, pp. 176–187. Springer, Cham (2019). https://doi.org/10.1007/978-3-030-27629-4_18
25. Thusoo, A., et al.: Hive-a petabyte scale data warehouse using hadoop. In: 2010 IEEE 26th International Conference on Data Engineering (ICDE 2010) (2010)
26. Tzitzikas, Y., et al.: Integrating heterogeneous and distributed information about marine species through a top level ontology. In: Garoufallou, E., Greenberg, J. (eds.) MTSR 2013. CCIS, vol. 390, pp. 289–301. Springer, Cham (2013). https://doi.org/10.1007/978-3-319-03437-9_29
27. Wang, K., Xu, G., Su, Z., Liu, Y.D.: GraphQ: graph query processing with abstraction refinement-scalable and programmable analytics over very large graphs on a single {PC}. In: 2015 Annual Technical Conference 2015 (2015)
28. Zapilko, B., Mathiak, B.: Performing statistical methods on linked data. In: International Conference on Dublin Core and Metadata Applications (2011)
29. Zhao, P., Li, X., Xin, D., Han, J.: Graph cube: on warehousing and OLAP multidimensional networks. In: Proceedings of the 2011 ACM SIGMOD International Conference on Management of Data (2011)

Incremental Evaluation of Continuous Analytic Queries in HIFUN

Petros Zervoudakis[1](✉), Haridimos Kondylakis[1], Dimitris Plexousakis[1],
and Nicolas Spyratos[2]

[1] Institute of Computer Science, FORTH, Heraklion, Greece
{zervoudak,kondylak,dp}@ics.forth.gr
[2] Laboratoire de Recherche en Informatique, UMR8623 of CNRS, Universite Paris-Sud 11,
Orsay, France
Nicolas.Spyratos@lri.fr

Abstract. A huge amount of data is generated each day from various sources.
Analysis of these massive data is difficult, and requires new forms of processing
to enable enhanced decision making, insight discovery and process optimization.
In addition, besides their ever increasing volume, datasets change frequently, and
as such, results to continuous queries have to be updated at short intervals. In
this paper, we address the problem of evaluating continuous queries over big data
streams that are frequently updated, adopting HIFUN, a high-level query language
introduced recently. HIFUN offers a clear separation between the conceptual layer,
where analytic queries are defined independently of the nature and location of data,
and the physical layer where queries are evaluated, by encoding them as map-
reduce jobs or as SQL group-by queries. Using HIFUN, we devise an algorithm
for incremental processing of continuous queries, processing only the most recent
data partition, and exploiting already computed information, without requiring
evaluating the query over the complete dataset. Subsequently, we translate the
generic algorithm to both SQL and MapReduce using SPARK, exploiting the
query rewriting method provided by HIFUN. The experiments performed show
the advantages of our solution in terms of query answering efficiency.

Keywords: Big data · Data analytics · Incremental processing · Query language

1 Introduction

Data emanating from high-speed streams is progressively prevalent in today's data
ecosystem. Example data streams, that are rapidly updated, include bank transactions,
network traffic data, IoT data, the Linked Open Cloud [1, 2] and so on. In order to extract
knowledge, find useful patterns, and act on information present in these streams, the data
need to be rapidly analyzed and processed. However, this is a challenge, when new data
arrive continuously at high speed, and efficient data processing algorithms are needed.

The research community has already provided open-source distributed batch processing systems like Hadoop [3] and MapReduce [4], that allow query processing over
static and historical datasets, enabling scalable parallel analytics. To this direction, Spark

© Springer Nature Switzerland AG 2020
G. Flouris et al. (Eds.): ISIP 2019, CCIS 1197, pp. 53–67, 2020.
https://doi.org/10.1007/978-3-030-44900-1_4

[5] has emerged on top of Hadoop, and has gained traction, with many advantages such as fault tolerance and efficient data processing - exploiting main memory storage. However, even with those technologies, processing and analyzing big volumes of data is not efficient enough, in scenarios that need rapid response to change over continuous data streams [6]. Consequently, a large amount of research works focus on stream processing, developing streaming engines such as Spark Streaming [7], Spark Structured Streaming [8], Storm [9], Flink [10], and Google Data Flow [11]. Processing of continuous queries is a major challenge in a streaming context. A continuous query is a query which is evaluated automatically and periodically over a dataset that changes over time [12]. The results of continuous queries are usually fed to dashboards, in large enterprises, to provide support in the decision-making process.

As new data and updates arrive at a high rate, query re-evaluation from scratch can incur significant delays. Therefore the problem is how to evaluate the query incrementally, that is, given the answer of the query at time t, on dataset D, how to find the answer of the query at time t' on dataset D', assuming that the answer at time t has been saved and results become stale and stagnant over time. Incremental processing is an auspicious approach for refreshing mining results as it uses previously saved results, to avoid the cost of re-computation from scratch. There is an obvious relationship between continuous queries and materialized views [13], since a materialized view is a derived database relation whose contents are periodically updated by either a complete or incremental refresh. Incremental view maintenance methods [14, 15] exploit differential algorithms to re-evaluate the view expression in order to enable the incremental update of materialized views.

In our case, we study this problem in the context of HIFUN, a recently proposed high level functional language of analytic queries [16, 17]. Two distinctive features of HIFUN are that (a) analytic queries and their answers are defined and studied in the abstract, independently of the structure and location of the data and (b) each HIFUN query can be mapped either to a map-reduce job or to an SQL group-by query. Our approach exploits both the Spark Streaming and the Spark Structured Streaming in the physical level to implement an incremental evaluation algorithm using HIFUN semantics. More specifically our contributions in this paper are the following: (a) we use the HIFUN language to define the continuous query problem in the abstract and give a generic algorithm for its solution, (b) we translate the generic algorithm to both SQL and MapReduce and (c) we implement the generic algorithm in SPARK (both, SQL and MapReduce) using the query rewriting method provided by HIFUN.

The experimental results indicate that, in terms of performance, our implementation is at least as good as the conventional ones. To the best of our knowledge our approach is unique in presenting incremental algorithms for both the high-level HIFUN language and the corresponding low-level mapping of those algorithms to the map-reduce and the group-by SQL models. The remaining of the paper is organized as follows. In Sect. 2, we present the theoretical framework and the query language model used. Then in Sect. 3, we describe our algorithms for incremental evaluation of continuous queries at the HIFUN level and in Sect. 4 the corresponding implementation at the physical level. In Sect. 5, we evaluate our system and, finally, in Sect. 6 we conclude the paper and discuss possible directions for future work.

2   The Query Language Model

In this section, we describe briefly the conceptual model of the HIFUN language. The model offers a clear separation between the conceptual and the physical level, which means that it can be used to define analytic queries and their evaluation independently of the specific nature and location of the data sets (structured, unstructured, centrally stored or distributed). For more details on the HIFUN language the interested reader is referred to the relevant papers [16, 17].

Analysis Context. The basic notion used in HIFUN is that of attribute of a dataset. In HIFUN, an attribute is seen as a function from the dataset to some domain of values. For example, if the dataset D is a set of tweets, then the attribute "character count" (denoted as cc) is seen as a functions $cc: D \rightarrow Count$ such that, for each tweet t, $cc(t)$ is the number of characters in t.

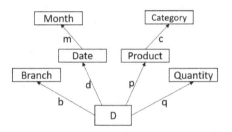

Fig. 1. Analysis context example

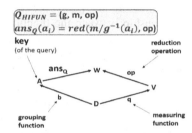

Fig. 2. A query Q and its answer ans_Q

Let us see an example to motivate the definition of a HIFUN query. Consider a distribution center (e.g. Walmart) which delivers products of various types in a number of branches and suppose D is the set of all delivery invoices collected over a year. Each delivery invoice has an identifier (e.g. an integer) and shows the branch in which the delivery took place, the date of delivery, the type of product delivered (e.g. CocaLight) and the quantity (i.e. the number of units delivered of that type of product). There is a separate invoice for each type of product delivered; and the data on all invoices during the year are stored in a data warehouse for analysis purposes. The information provided by each invoice would most likely be represented as a record with the following fields: Invoice number, Branch, Date, Product, Quantity. In the HIFUN approach, this information is seen as a set of four functions, namely d, b, p and q, as shown in Fig. 1, where D stands for the set of all invoice numbers and the arrows represent attributes of D. Following this view, given an invoice number, the function d returns a date, the function b a branch, the function p a product type and the function q a quantity (i.e. the number of units of that product type). The attributes d, b, p and q of our running example are "direct" attributes of D in the sense that their values appear on the delivery invoices. However, apart from these attributes, analysts might be interested in attributes that are not direct but can be "derived" from the direct attributes. Figure 1 also shows the direct attributes with several derived attributes: attribute m can be derived from attribute d (e.g. from the date 24/10/1992 one can derive the month 10/1992); and attribute c can be derived from a product master table. The set of all attributes (direct and derived) that are of interest to a group of analysts is called an analysis context (or simply a context).

Query Definition. A query is defined to be an ordered triple $Q = (g, m, op)$ such that g and m are attributes of the dataset D, and op is an aggregate operation on m-values. The attributes g and m are called *"grouping attributes"* and *"measuring attributes"* respectively. Formally, we have the following definition: let D be a finite set of data items, such that $D = \{d_1, \ldots, d_n\}$. An analytic HIFUN query over D is an ordered triple $Q = (g, m, op)$, where g is a function with domain the set D and range a set A, m is a function with domain the set D and range a set V, and op is an operation over V taking its values in a set W. If $\{a_1, \ldots, a_n\}$ is a set containing the values of g over D (clearly $k <= n$), then we call grouping of D by g, the partition $\pi_g = \{g^{-1}(a_1), \ldots, g^{-1}(a_k)\}$ induced by g on D. The reduction of m with respect to op, denoted $red(m, op)$ is a value of W defined as $red(m, op) = op(< m(d_1), \ldots, m(d_n) >)$. On the basis of the above definitions, the answer to Q, denoted as ans_Q, is a function from the set of values of g to W defined by $ans_Q(a_i) = red(m/g^{-1}(a_i), op)$, $i = 1, 2, \ldots, k$. Figure 2 shows the relationship between the function ans_Q and the functions appearing in the query $Q = (g, m, op)$.

A query $Q = (g, m, op)$ over D can be enriched by introduction functional restriction at either of two levels: at the level of attributes or at the level of the query answer. An *attributed-restricted* query is defined as $Q = (g/E, m, op)$, where E is any subset of D. It is evaluated by computing the restriction g/E and then evaluating the query $(g/E, m, op)$ over E. A *result-restricted* query is defined as $Q = (g, m, op)/F$, where F is any subset of the target of the domain of definition of ans_Q. It is evaluated by evaluating the query $Q = (g, m, op)$ over D, to obtain its answer ans_Q, and then computing the restriction ans_Q/F.

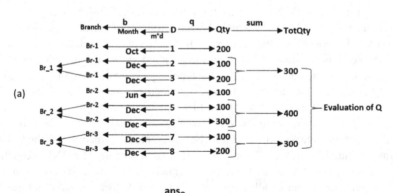

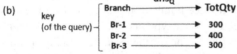

Fig. 3. An analytic query and its answer

Analysts can express analytic queries within their context by defining triples of the form (g, m, op), where g and m are attributes of any node of the context. Also, complex grouping functions can be defined, using the following three functional algebra operations: *composition(○), pairing (^), restriction(/)* and *Cartesian product projection.*

Returning to our running example, assume that we want to know the total quantity delivered to each branch only for month '*December*'. Formally, this query is written as $Q = (b/E, q, sum)$, where $E = \{x | x \in D^\wedge (m°d)(x) = \text{'December'}\}$. This computation needs only three functions, namely b, q and $m°d$ among the set of functions that are defined in context of Fig. 1. Figure 3(a) illustrates an example of the data returned by b, q and $m°d$ and the computations needed during the query evaluation process. In order to find the total quantity by branch for month "*December*", the following steps should be executed: *(a) Grouping:* The grouping based on b/E creates a group for each branch which is different than the obtained when grouping is based on b. During this step, all invoices that happened in month '*December*, referring to the same branch are grouped together; *(b) Measuring:* In each group computed during the previous step, we find the quantity corresponding to each invoice by extracting the value using the function q. *(c) Reduction:* For each group, we sum up the quantities. Then the relation of each branch to the corresponding total quantity is the evaluation of query Q, illustrated in Fig. 3(b).

Query Rewriting. The formal model of HIFUN supports also query rewriting. An incoming query can be rewritten to other queries possibly reducing evaluation cost, based on the basic idea that a functional expression, when used as a grouping function, can be equivalently rewritten to other expressions. This observation leads to our basic rewriting rule for queries that have a common measuring function and operation but different grouping functions and require that the aggregate operation to be distributive. To see intuitively how the basic rewriting rule works, consider the following queries on the context of Fig. 1. The query $Q = (p, q, sum)$ asking the totals by product and the query $Q' = (c°p, q, sum)$ asking for the totals by category. Clearly, the query Q' can be answered directly, following the abstract definition of answer (i.e. by grouping, measuring and reduction). However, Q' can also be answered, if we know (a) the totals by product and (b) which products are in which category. Then all we have to do is to sum up the totals by product in each category to find the totals by category. Now, the totals by product are given by the answer to Q, and the association of products with categories is given by the function c. Therefore, the query Q' can be answered by the following query Q'', which uses the answer of Q as its measure : $Q'' = (c, ans_Q, sum)$, asking for the sum of product totals by category. Note that the query Q'' is well formed as c and ans_Q have Product as their (common) source. Besides the basic rewriting rule, other such rules are available [17].

Conceptual Query Evaluation Scheme. Using the batch processing approach we first have to store the available data and then evaluate the query. In detail, the following steps have to be followed:

a. *Query Input Preparation. IN(Q)* denotes the set of tuples which contain the information for evaluating query Q, independently of whether the dataset is centrally or distributed stored. In this step k sets of tuples $I_1, \ldots, I_k$ are returned, that form a partition $\pi_{IN(Q)}$ of the input $IN(Q)$, where each tuple contains a data item identifier and the values of its attributes g and m, including the values of any possible attributes contained in the query restrictions.

b. *Attribute Filtering.* If there are no attribute restrictions on query definition, this step is skipped. Elsewhere, filtering is performed on *IN(Q)* tuples according to the query attribute restrictions.

c. π_g *Construction.* This step constructs the partition $\pi_g = \{G_1, \ldots, G_n\}$, as it was previously defined in the query definition. The reduction of π_g will produce the answer to the query.

d. π_g *Reduction.* Once the block G_j has been constructed, it can be reduced by the operation defined in the query definition, to obtain the answer on the value g_j of **g:**
$\text{ans}_Q(g_i) = \text{red } (m/G_j, op)$.

e. *Result Filtering.* If there are no result restrictions on query definition, this step is skipped. Elsewhere filtering is performed on **ans_Q** according the restriction on the query results.

In our running example, the query $Q = (b/E, q, sum)$, where $E = \{x \mid x \in D^\wedge (m^\circ d)(x) = \text{'December'}\}$ is mapped to the aforementioned conceptual schema as illustrated in Fig. 4.

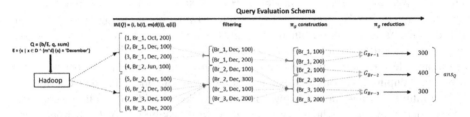

Fig. 4. The conceptual schema steps

3   Incremental Computation

In this section, we show how we can use the HIFUN language to incrementally evaluate continuous queries. An important common feature of real-life applications is that the input data continuously grow and old data remain intact. As such, for the rest of this paper we assume that the dataset being processed can only increase in size between successive time moments t and t'. In such a scenario, the idea of incremental computation of a continuous query is to use the results of an already performed computation on old data and evaluate the query only on the lately appended data, merging eventually new and previous results.

Figure 5 illustrates our proposed incremental approach for continuous queries – the same query asked two times. We perceive the problem of incremental evaluation as follows: given the answer of a query Q at time t, on dataset D, find the answer of the query at time t' on data set D', where $D' = D + \Delta D$, by evaluating the query only on ΔD and reusing the answer on D.

Now assume that the function ans_Q is the answer on D of Q at time t, including K groups of answers, and that the function $incr_Q$ is the answer on $\Delta D = D'/D$ of Q at

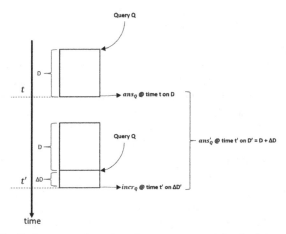

Fig. 5. Incremental computation over an append-only data set.

time t', including the K' groups of answers. If the reduction operation *op* is a *distributive operation*, the answer *ans'* of query Q at time t', is evaluated as follows:

- **op = sum:** $ans'(i) = ans(i) + incr(i)$ *if i is in* $K \cap K'$;
  $ans(i)$ *if i is in* $K \backslash K'$; $incr(i)$ *if i is in* $K' \backslash K$
- **op = min:** $ans'(i) = min(ans(i), incr(i))$ *if i is in* $K \cap K'$;
  $ans(i)$ *if i is in* $K \backslash K'$; $incr(i)$ *if i is in* $K' \backslash K$
- **op = max:** $ans'(i) = max(ans(i), incr(i))$ *if i is in* $K \cap K'$;
  $ans(i)$ *if i is in* $K \backslash K'$; $incr(i)$ *if i is in* $K' \backslash K$
- **op = count:** $ans'(i) = ans(i) + incr(i)$ *if i is in* $K \cap K'$;
  $ans(i)$ *if i is in* $K \backslash K'$; $incr(i)$ *if i is in* $K' \backslash K$

Aggregate operations operate on a set of values to compute a single value as a result [18]. Distributive aggregate operations are those whose computation can be *"distributed"* and be recombined using the distributed aggregates. All the operations that are previously described are distributive. This means that if the data are distributed into n sets, and we apply the aforementioned distributive operation to each one of them (resulting in n aggregate values), the total aggregate operation can be computed for all data by applying the aggregate operation for each subset and then combining the results. For example: *sum (1, 2, 3, 4, 5) = sum (sum (1, 2), sum (3, 4, 5))*.

We also support non-distributive aggregate operations such as the average as: *avg (1, 2, 3, 4, 5) ≠ avg (avg (1, 2), avg (3, 4, 5))*. Non-distributive aggregate operations can be computed by algebraic functions that are obtained by applying a combination of distributive aggregate functions. For example, the average can be computed by summing a group of numbers and then dividing by the count of those numbers. Both, sum and count are distributive operations. More specifically:

- **op = avg:** $ans'(i) = ans(i)$ *if i is in* $K \backslash K'$; $ans'(i) = incr(i)$ *if i is in* $K' \backslash K$

$$\text{ans}'(i) = \frac{ans_{op=sum}(i)+incr_{op=sum}(i)}{ans_{op=count}(i)+incr_{op=count}(i)} \; if \; i \; is \; in \; K \cap K'$$

Finally, there are additional aggregate operations, whose computation requires looking at all the data at once, and hence their evaluation cannot be decomposed into smaller pieces. Common examples of this type of aggregate operations include median and count-distinct. However, we leave those operations for future work.

Now consider the example illustrated in Fig. 6. We would like to know the total quantity delivered to each branch during the month *December*. At time t the query was evaluated over the dataset D, returning the function ans_Q: $Branch \rightarrow TotQty$, as the answer of Q. Then, at time t' the query was again evaluated over only the dataset ΔD, returning the function $incr_Q$: $Branch \rightarrow TotQty$, as the answer of Q on ΔD. In this case, the aggregate operation is the distributive operation sum. As such, we can produce the ans'_Q on time t' merging the functions ans_Q and $incr_Q$ as follows: The groups that appear only in K, which are the groups returned by the query Q at time t on D, are transferred directly to the result of ans'_Q. The groups that appear only in K', which are the groups of the query Q at time t' on ΔD, are transferred directly to the result of ans'_Q. The distributive operation sum is applied when the groups appear in the intersection of K and K'. For example, the key Br-2 appears in both K and K', therefore the answer ans'_Q for that key resulting as $sum \; (400 + 200) = 600$.

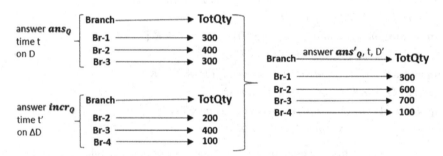

Fig. 6. Incremental evaluation on our running example.

As already mentioned, HIFUN offers query rewriting which is possible to reduce the evaluation cost. Assume for example the context of Fig. 1 and the rewritten query $Q = (c, (p, q, sum), sum)$. Assume also that the rewritten query Q has already been evaluated on D at time t and the function ans_Q: $Category \rightarrow Totals$ is the answer of Q. Figure 7 shows how we leverage the basic rewriting rule, to evaluate the query Q on ΔD at time t'. The rewriting rule requires the evaluation of the base query $Q_{base} = (p, m, sum)$ only on ΔD at time t'. The query Q_{base}. is executed and the answer is returned as $ans_{Q_{base}}$: $Product \rightarrow TotQty$. Therefore, the query Q can be answered on ΔD at time t' by evaluating the following query $Q' = (c, ans_{Q_{base}}, sum)$. The answer of the rewritten query Q', (the equivalent query of Q on ΔD) is computed by combining the function $incr_Q$ on ΔD at time t' and the function ans_Q on D at time t as previously described.

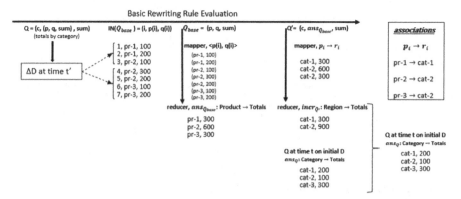

Fig. 7. Example of rewritten query

4 System Implementation

As already shown, HIFUN queries can be defined at the conceptual level independent of the nature and the location of the data. These queries can be evaluated by encoding them either as map-reduce jobs or SQL group-by queries, depending on the nature of the available data. In this section, we show how to physically evaluate a HIFUN query processing live data streams. This is implemented using two different physical layer mechanisms: (1) the Spark Streaming [6] and (2) the Spark Structured Streaming [7]. Both mechanisms support the micro-batching concept - fragmentation of the stream as a sequence of small batch chunks of data. On small intervals, the incoming stream is packed to a chunk of data and is delivered to the system to be further processed [18].

4.1 Micro-batch Stream Processing

In the micro-batching approach, as a dataset continuously grows, and as new data become available, we process the tuples in discrete batches. The batches are processed sequentially and as a high volume of tuples can be processed per micro batch, the aforementioned mechanism uses parallelization to speed up data processing. As such, we assume an initial dataset D_i which is followed by a continuous stream of incremental batches ΔD_i that arrive at consecutive time intervals Δt. As we already explained, incremental evaluation would produce the query results at time $t + \Delta t$ by simply combing the query results at time t, with the results from processing the incremental batches ΔD_i. Two key observations should be made here. The first is that computations needed are solely performed within the specific batch, following the evaluation scheme described in the previous section. Therefore, for every batch interval we calculate a result based on delta subset ΔD_i, e.g. $incr_i \leftarrow e(\Delta D_i)$. The second observation is that a state should be kept across all batches. Stateful processing is able to handle unbounded streams of data. After the evaluation of each query is completed for each micro-batch, we need to keep the state across all batches. The previous state value and the current delta result are merged together and the system produces a new state incrementally, e.g. $state \leftarrow u(incr_i, state)$. Figure 8, illustrates this incremental approach.

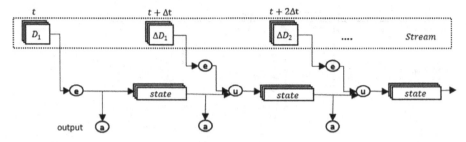

Fig. 8. State maintenance.

4.2 Continuous HIFUN Query to MapReduce

Our conceptual evaluation scheme is first implemented using the Map-Reduce programming model over the physical layer, exploiting Spark Streaming. Spark Streaming is a
stream processing framework based on the concept of discretized streams and provides
the *DStream API* which accepts sequences of data arriving over time. The API implements the micro-batch stream processing approach with periodic checking of internal
state at each batch interval. Internally, each DStream item is a sequence of data called
Resilient Distributed Datasets (RDDs), kept in main memory. The batch interval value
can also be set to specify how often an input RDD is generated. In the following, we
describe in detail the mapping of our conceptual schema earlier presented for incremental
computation to the physical layer.

4.3 Conceptual Schema to MapReduce

In this section we elaborate on the generic query evaluation scheme, described in Sect. 2,
presenting details on its implementation over the physical layer:

a. *Query Input Preparation.* In this step, a set of attributes - which are included in
   grouping and measuring part of Q - are used to extract the necessary information
   from the unstructured dataset received as input. Then, the $IN(Q)$ set is computed
   based on the identified attributes for each record. To this propose, we iterate over all
   input records of the DStream and return a new version which contains preprocessed
   information for the next evaluation steps.
b. *Attributes filtering.* If attribute restrictions exist, this step filters out the corresponding DStream tuples. The filter method returns a new DStream containing only the
   elements that satisfy those restrictions.
c. π_g *construction:* To construct the grouping part π_g, each mapper, receives the tuples
   to be used for extracting the key-values pairs from each data item. The result of this
   step, is a new PairDStream which contains key-value pairs (K, V). The key K is
   the value of the grouping attribute of each data item, or the value of the grouping
   attributes, if the domain of ans_Q is a cartesian product of two or more grouping
   attributes. The value V is the value of the measuring attribute of each data item.
d. π_g *reduction.* In this step, each reducer uses the query operation op to reduce the
   set of key-value pairs received. The reduce-by-key method is applied and a new
   DStream is returned.

e. *Result filtering.* If result restrictions exist, this step filters out the tuples of the DStream that don't conform to the query restrictions. The filter method is applied on DStream and a new DStream is returned containing only the elements that satisfy the queried predicates.

Assume now the following HIFUN query $Q = (k, u, op)$, where k and u are the functions used by the mappers to extract the key-value pairs during the input preparation step, and op is the operation applied by the reducers. If $k = g \,°f$ is the composition of the two functions, then the query Q can be rewritten under the basic rewriting rule as follows: $Q' = (g, (f, u, op), op)$. This implies that the initial query Q can be rewritten as a sequence of the two other queries. The base query $Q_{base} = (f, u, op)$ should be executed first. The resulting query $Q' = \left(g, ans_{Q_{base}}, op\right)$ should then be executed as follows: the mapper used to construct the key-values pairs is using the association of f with g, that is provided by the function g, and then, the reducer applies the reduction by the operation op on the set of constructed key-pairs.

Independent of whether Q is evaluated as is, or if is rewritten under the basic rewriting rule, the produced answer is a function from a domain of values to a set of values. The domain of values is a set of keys, each of those correlated with the key of the query. The incremental algorithm examines the set of keys independently of whether those keys occurred after evaluating the original query Q or the rewritten one. In the next subsection, the details of the incremental evaluation are provided.

4.3.1 Incremental Evaluation
The aforementioned jobs are executed using Spark Streaming for each incoming micro-batch. When a query is executed, an answer is produced for a micro-batch and a DStream is created, encapsulating a key-value pair in the form of a *DStream [(K, V)]*, where k is the key of the continuous query that appears in the current micro-batch and V is the value of the reduction operation. We have to note that we maintain the state across the micro-batches (using the mapWithState method), using the key-value pairs produced from each micro-batch. As such, we are able to execute partial updates for only the newly arrived keys in the current micro-batch, initiating the computations only for the records that need to be updated. The state information is stored as a mapWithStateRDD, thus benefiting from the distribution's efficiency and effectiveness of Spark.

4.4   Translating Continuous HIFUN Queries to SQL

In this section, we explain how a HIFUN query can be evaluated. We describe how we map the conceptual evaluation schema to the existing physical level mechanism using the semantics of the SQL exploiting group-by SQL queries of Spark Structured Streaming. The basic idea in Structured Streaming is treating continuously arriving data, as a table, that is being continuously appended. Structured Streaming runs in a micro batch execution model as well. Spark waits for a time interval and batches together all events that were received during that interval. The mapping mechanism defines a query on the input table, as if it was a static table, computing a result table that will be updated through the data stream. Spark automatically converts this batch-like query to a streaming execution plan. This is called instrumentalization: Spark figures out what needs to be maintained to update the result each time a new batch arrives. At each time interval, Spark checks for new rows in the input table and incrementally updates the result. As soon as a micro-batch execution is complete, the next batch is collected and the process is reapplied.

In [13] and [16] is already shown how HIFUN queries can be mapped to SQL group-by queries. To evaluate the attribute- and result restricted HIFUN query $Q = (g_A/E, m_B, op)/F$, where $E = \{x \text{ in } D/g_A (x) = \text{'ABC'}\}$ and $F = \{y \text{ in } A/ans_Q(y) < 123\}$, assume that the attributes A and B appears in the same table T. Figure 9 shows the correspondence between our conceptual evaluation scheme and evaluation by a group-by sql query. For computing queries on streaming data we apply similar techniques to batch computation on static data. The Spark SQL engine executes this incrementally and continuously updates the result as streaming data arrive.

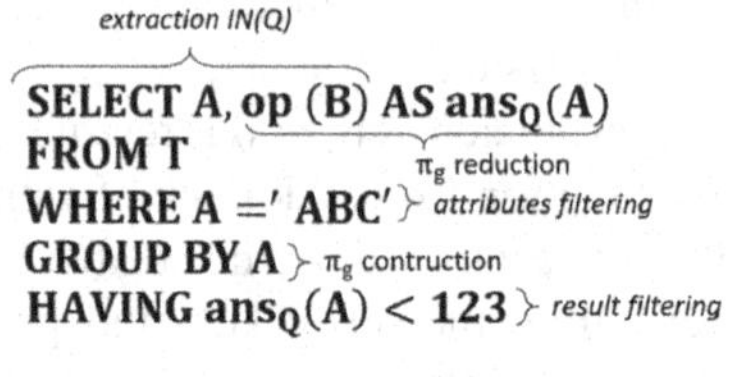

extraction IN(Q)

$\overbrace{\text{SELECT A, op (B)} \text{ AS ans}_Q(A)}$
FROM T π_g reduction
WHERE A $=' $ **ABC'** $\}$ attributes filtering
GROUP BY A $\}$ π_g contruction
HAVING ans$_Q$(A) $<$ 123 $\}$ result filtering

Fig. 9. The group-by SQL query decomposed into steps of our evaluation schema.

For executing the rewritten query $Q = (g, (f, u, op), op)$ using SQL, the base query $Q_{base} = (f, u, op)$ should be first executed, using the mapping of the Q_{base} to the corresponding SQL group-by query, producing the intermediate table. To evaluate the result query (g, ans_{base}, op), the intermediate table is joined with the table that contains the grouping attribute g, and the aggregation function is applied on the column that contains the result of the aggregation of the base query.

5 Experimental Results

We expect that implementing an incremental query evaluation mechanism will result in a significant improvement to the overall query evaluation performance. In the following experiments, we compare our incremental approach with the batch processing approach and we present the effectiveness of the query rewritings.

To perform our evaluation we generated a synthetic dataset, using the analysis contexts shown in Fig. 10. In the case of the MapReduce execution model, the source dataset is provided as a single text file, whereas in the case of the SQL execution model, we transform the dataset into the corresponding relational schema.

In order to evaluate the effectiveness of the incremental evaluation of a query, we define the following query $Q = (g_1, m, sum)$.. We begin our experiments with an initial dataset of 80M records. That dataset continuously grows over time and in each batch, 80M new records are added to the existing dataset. Using this dataset, *the batch computation approach* looks at the entire dataset each time as new data is available to be processed. The incremental approach on the other hand, only examines the new incoming

Fig. 10. Analysis context of the unstructured and structured synthetic dataset.

data and incorporates the increment in the result. Figure 11 shows the performance of the two approaches when the HIFUN query is evaluated using MapReduce jobs or group-by SQL queries. The results show that using the incremental approach we gain a great benefit: while the dataset grows over a time, the evaluation cost remains stable independent of the overall increase in data size. In contrast, when the queries are evaluated over the whole batch of data, the evaluation cost increases as the size of the input batch data increases as well.

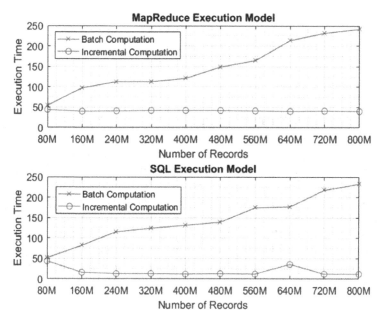

Fig. 11. Evaluation of continuous HIFUN query

Now we present another experiment on the unstructured increasing dataset described by the context of Fig. 10. We define the following set of queries $Q = \left\{ \left(g_{11}^{\circ} g_1, m, sum \right), \ldots, \left(g_{15}^{\circ} g_1, m, sum \right) \right\}$ containing five queries, all of them having the same distributive operation applicable on the same measuring attribute m. As described in the previous sections, Q can be equivalently rewritten using the basic rewriting rule as follows $Q' = \{ (g_{11}, (g_1, m, op), sum), \ldots, (g_{15}, (g_1, m, op), sum) \}$. The rewritten set Q' consists of five queries and each one uses the answer of (g_1, m, op) as its measure.

To investigate the effectiveness of the rewriting rule, we run the experiments for a set Q' starting initially with one query and adding each time another one, till we reach a total number of five queries to be executed each time. We notice that using the basic rewriting rule we have real efficiency benefits, as more queries participate in the rewritten Q'. Table 1 shows the average evaluation time of the rewritten set Q' presenting the number of the queries issued each time. The *average evaluation time* is the average time to evaluate incrementally the set of the continuous queries over our synthetic dataset scenario. The result shows that although the number of the queries that should be answered each time increases, the average time for evaluation those queries is stable.

Table 1. Basic rewriting rule evaluation

Num. of queries	Avg. Eval. Time
1	38.9 s
2	39.5 s
3	40.1 s
4	40.2 s
5	40.5 s

6 Conclusions and Future Work

In this paper, we leverage the HIFUN language, adding an incremental evaluation mechanism using Spark Streaming. We present an approach allowing the incremental update of continuous query results, preventing the costly re-computation from scratch. We showed also that query rewriting, enabled by the adoption of the HIFUN language, can be implemented in the physical layer as well, further benefiting the efficiency of query answering. We demonstrated experimentally the considerable advantages gained by using the incremental evaluation, reducing the overall evaluation cost using both the map-reduce implementation and the SQL one. Future work will exploit more complex rewritings to the physical layer, further minimizing the evaluation cost sets of queries – where multiple queries could benefit from intermediate results.

References

1. Agathangelos, G., Troullinou, G., Kondylakis, H., Stefanidis, K., Plexousakis, D.: Incremental data partitioning of RDF Data in SPARK. In: Gangemi, A., et al. (eds.) ESWC 2018. LNCS, vol. 11155, pp. 50–54. Springer, Cham (2018). https://doi.org/10.1007/978-3-319-98192-5_10
2. Agathangelos, G., Troullinou, G., Kondylakis, H., et al.: RDF Query answering using apache spark: review and assessment. In: ICDE Workshops, pp. 54–59 (2018)
3. White, T.: Hadoop: The Definitive Guide. O'Reilly Media, Inc., Sebastopol (2009)

4. Dean, J., Ghemawat, S.: MapReduce: simplified data processing on large clusters. Commun. ACM **51**, 107–113 (2004)
5. Zaharia, M.A., Chowdhury, M., Franklin, M.J., Shenker, S., Stoica, I.: Spark: cluster computing with working sets. Ann. Emerg. Med. **39**(6), 691–692 (2002)
6. Karimov, J., Rabl, T., Katsifodimos, A., Samarev, R., Heiskanen, H., Markl, V.: Benchmarking distributed stream data processing systems. In: 2018 IEEE 34th International Conference on Data Engineering (ICDE), pp. 1507–1518 (2018). Author, F.: Contribution title. In: 9th International Proceedings on Proceedings, pp. 1–2. Publisher, Location (2010)
7. Zaharia, M.A., Das, T., Li, D.H., Hunter, T., Shenker, S., Stoica, I.: Discretized streams: fault-tolerant streaming computation at scale. In: SOSP (2013)
8. Armbrust, M., et al.: Structured streaming: a declarative API for real-time applications in apache spark. In: SIGMOD Conference (2018)
9. Iqbal, M.S., Soomro, T.R.: Big data analysis: apache storm perspective. Int. J. Comput. Trends Technol. **19**, 9–14 (2015)
10. Carbone, P., Katsifodimos, A., Ewen, S., Markl, V., Haridi, S., Tzoumas, K.: Apache Flink™: stream and batch processing in a single engine. IEEE Data Eng. Bull. **38**, 28–38 (2015)
11. Akidau, T., et al.: The dataflow model: a practical approach to balancing correctness, latency, and cost in massive-scale, unbounded, out-of-order data processing. PVLDB **8**, 1792–1803 (2015)
12. Babu, S., Widom, J.: Continuous queries over data streams. ACM SIGMOD Rec. **30**, 109–120 (2001)
13. Gupta, A., Mumick, I.S.: Materialized Views: Techniques, Implementations, and Applications. MIT Press, Cambridge (1999)
14. Blakeley, J.A., Larson, P., Tompa, F.W.: Efficiently updating materialized views. ACM SIGMOD Rec. **15**, 61–71 (1986)
15. Ahmad, Y., Kennedy, O., Koch, C., Nikolic, M.: DBToaster: higher-order delta processing for dynamic, frequently fresh views. PVLDB **5**, 968–979 (2012)
16. Spyratos, N., Sugibuchi, T.: HIFUN - a high level functional query language for big data analytics. J. Intell. Inf. Syst. **51**, 529–555 (2018). https://doi.org/10.1007/s10844-018-0495-6
17. Spyratos, N., Sugibuchi, T.: A high-level query language for big data analytics (2014)
18. Jesus, P., Baquero, C., Almeida, P.S.: A survey of distributed data aggregation algorithms. IEEE Commun. Surv. Tutorials **17**, 381–404 (2011)

Evolution of a Data Series Index

The iSAX Family of Data Series Indexes: iSAX, iSAX2.0, iSAX2+, ADS, ADS+, ADS-Full, ParIS, ParIS+, MESSI, DPiSAX, ULISSE, Coconut-Trie/Tree, Coconut-LSM

Themis Palpanas[✉]

University of Paris, Paris, France
`themis@mi.parisdescartes.fr`

Abstract. There is an increasingly pressing need, by several applications in diverse domains, for developing techniques able to index and mine very large collections of sequences, or data series. It is not unusual for these applications to involve numbers of data series in the order of billions, which are often times not analyzed in their full detail due to their sheer size. In this work, we describe techniques for indexing and efficient similarity search in truly massive collections of data series, focusing on the iSAX family of data series indexes. We present their design characteristics, and describe their evolution to address different needs: bulk loading, adaptive indexing, parallelism and distribution, variable-length query answering, and bottom-up indexing. Based on this discussion, we conclude by presenting promising research directions.

Keywords: Data series · Time series · Sequences · Indexing · Analytics

1 Introduction

Data series have gathered the attention of the data management community for almost three decades [54], and still represent an active and challenging research direction [7,56,83]. Data series are one of the most common data types, present in virtually every scientific and social domain [56]: they appear as audio sequences [34], shape and image data [76], financial [67], environmental monitoring [64], scientific data [30], and others. It is nowadays not unusual for applications to involve numbers of sequences in the order of billions [1,2].

A *data series*, or *data sequence*, is an ordered sequence of data points[1]. Formally, a data series $T = (p_1, \dots p_n)$ is defined as a sequence of points $p_i = (v_i, t_i)$, where each point is associated with a value v_i and a time t_i in which this recording was made, and n is the size (or length) of the series. If the dimension that imposes the ordering of the sequence is time then we talk about *time series*,

---

[1] For the rest of this paper, we are going to use the terms *data series* and *sequence* interchangeably.

© Springer Nature Switzerland AG 2020
G. Flouris et al. (Eds.): ISIP 2019, CCIS 1197, pp. 68–83, 2020.
https://doi.org/10.1007/978-3-030-44900-1_5

though, a series can also be defined over other measures (e.g., angle in radial profiles in astronomy, mass in mass spectroscopy, position in biology, etc.).

A key observation is that analysts need to process and analyze a sequence (or subsequence) of values as a single object, rather than the individual points independently, which is what makes the management and analysis of data sequences a hard problem. In this context, Nearest Neighbor (NN) queries are of paramount importance, since they form the basis of virtually every data mining, or other complex analysis task involving data series [56]. However, NN queries on a large collection of data series are challenging, because data series collections grow very large in practice [13, 63]. Thus, methods for answering NN queries rely on two main techniques: data summarization and indexing. Data series summarization is used to reduce the dimensionality of the data series [3, 16, 35, 36, 43, 44, 62], and indexes are built on top of these summarizations [5, 62, 66, 70, 72].

In this study, we review the iSAX family of data series indexes, which all use the iSAX summarization technique to reduce the dimensionality of the original sequences. These indexes have attracted lots of attention, and represent the current state-of-the-art for several variations of the general problem. In particular, we present the iSAX summarization and discuss how it can be used to build the basic iSAX index [68, 69]. We describe iSAX2.0 [12] and iSAX2+ [13], the first data series indexes that inherently support bulk loading, allowing us to index datasets with 1 billion data series. We present the ADS and ADS+ indexes [78–80], which are the first adaptive data series indexes than can start answering queries correctly before the entire index has been built, as well as ADS-Full [80], which based on the same principles leads to an efficient 2-pass index creation strategy. We discuss ParIS [58] and ParIS+ [60], the first parallel data series indexes designed for modern hardware, and MESSI [59], a variation optimized for operation on memory-resident datasets. DPiSAX [42, 74, 75] is a distributed index that operates on top of Spark. We present ULISSE [45, 46], which is the first index that can inherently support queries of varying length. Finally, we describe Coconut [38–40], the first balanced index, which is built in a bottom-up fashion using a sortable iSAX-based summarization.

It is interesting to note that these indexes can be used not only for similarity search of data series, but also of general high-dimensional vectors [23, 24], leading to better performance than other high-dimensional techniques (including the popular LSH-based methods) [24].

By presenting all these indexes together[2], we contribute to the better understanding of the particular problems that each one solves, the way that their features could be combined, and the opportunities for future work.

2   Background and Preliminaries

[**Data Series Queries**]. Analysts need to perform (a) simple Selection-Projection-Transformation (SPT) queries, and (b) more complex Data-Mining

---

[2] More details on the topics of this paper can be found elsewhere [12, 13, 19, 20, 23, 27, 28, 38–40, 42, 45, 46, 48–51, 53–56, 58–60, 74, 75, 78, 79, 81–83].

(DM) queries. Simple SPT queries are those that select sequences and project points based on thresholds, point positions, or specific sequence properties (e.g., "above", "first 10 points", "peaks"), or queries that transform sequences using mathematical formulas (e.g., average). An example SPT query could be one that returns the first x points of all the sequences that have at least y points above a threshold. The majority of these queries could be handled (albeit not optimally) by current data management systems, which nevertheless, lack a domain specific query language that would support and facilitate such processing. DM queries on the other hand are more complex: they have to take into consideration the entire sequence, and treat it as a single object. Examples are: queries by content (range and similarity queries), clustering, classification, outlier, frequent sub-sequences, etc. These queries cannot be efficiently supported by current data management systems, since they require specialized data structures, algorithms, and storage methods.

Note that the data series datasets and queries may refer to either static, or streaming data. In the case of streaming data series, we are interested in the sub-sequences defined by a sliding window. The same is also true for static data series of very large size (e.g., an electroencephalogram, or a genome sequence), which we divide into sub-sequences using a sliding (or shifting) window. The length of these sub-sequences is chosen so that they contain the patterns of interest.

One of the most basic data mining tasks is that of finding similar data series, or NN in a database [3]. Similarity search is an integral part of most data mining procedures, such as clustering [73], classification and deviation detection [11,17]. Even though several distance measures have been proposed in the literature [6,10,18,21,51,71], the Euclidean distance is the most widely used and one of the most effective for large data series collections [22]. We note that an additional advantage of Euclidean distance is that in the case of Z-normalized series (mean $= 0$, stddev $= 1$), which are very often used in practice [81,82], it can be exploited to compute Pearson correlation [61].

[Data Series Summarizations]. A common approach for answering such queries is to perform a dimensionality reduction, or summarization technique. Several such summarizations have been proposed, such as the Discrete Fourier Transform (DFT) [3], the Discrete Wavelet Transform (DWT) [16], the Piecewise Aggregate Approximation (PAA) [36,77], the Adaptive Piecewise Constant Approximation (APCA) [15], or the Symbolic Aggregate approXimation (SAX) [44]. Note that that on average, there is little difference among these summarizations in terms of fidelity of approximation [22,57] (even though it *is* the case that certain representations favor particular data types, e.g., DFT for star-light-curves, APCA for bursty data, etc.).

These summarizations are usually accompanied by distance bounding functions that relate distances in the summarized space to distances in the original space through either lower or upper-bounding. With such bounding functions, we can index data series directly in the summarized space [5,62,66,70,72], and use these indexes to efficiently answer NN queries on large data series collections.

[Data Series Indexing]. Even though recent studies have shown that in certain cases sequential scans can be performed very efficiently [63], such techniques are only applicable when the database consists of a single, long data series, and queries are looking for potential matches in small subsequences of this long data series. Such approaches, however, do not bring benefit to the general case of querying a mixed database of several data series. Therefore, indexing is required in order to efficiently support data exploration tasks, which involve ad-hoc queries, i.e., the query workload is not known in advance.

A large set of indexing methods have been proposed for the different data series summarization methods, including traditional multidimensional [9,29,37,62] and specialized [5,66,70,72] indexes. Moreover, various distance measures have been presented that work on top of such indexes, e.g., Discrete Time Warping (DTW) and Euclidean Distance (ED).

Indexing can significantly reduce the time to answer DM queries. Nevertheless, recent studies have observed that the mere process of building the index can be prohibitively expensive in terms of time cost [12,13,78]: e.g., the process of creating the index for 1 billion data series takes several days to complete. This problem can be mitigated by the bulk loading technique. Bulk-loading has been studied in the context of traditional database indexes, such as B-trees and R-trees, and other multi-dimensional index structures [4,25,32,33,41,65].

3   The iSAX Family of Indexes

In this section, we describe the iSAX family of indexes, that is, all the indexes that are designed based on the iSAX summarization, and discuss their evolution over time. Figure 1 depicts the lineage of these indexes, along with the corresponding timeline. We note that all these indexes support both Z-normalized and non Z-normalized series, and the same index can answer queries using both the Euclidean and Dynamic Time Warping (DTW) distances (in the way mentioned in [59]), for k-NN and ϵ-range queries [23]. Finally, recent extensions of some of these indexes demonstrate that they can efficiently support approximate similarity search with quality guarantees (deterministic and probabilistic) [24], and that they dominate the state-of-the-art in the case of general high-dimensional vectors, as well [23,24].

3.1   The iSAX Summarization and Basic Index

The Piecewise Aggregate Approximation (PAA) [36,77] is a summarization technique that segments the data series in equal parts and calculates the average value for each segment. An example of a PAA representation can be seen in Fig. 2; in this case the original data series is divided into 4 equal parts. Based on PAA, Lin et al. [44] introduced the Symbolic Aggregate approXimation (SAX) representation that partitions the value space in segments of sizes that follow the normal distribution. Each PAA value can then be represented by a character (i.e., a small number of bits) that corresponds to the segment that it falls into.

This leads to a representation with a very small memory footprint, an important advantage when managing large sequence collections. A segmentation of size 3 can be seen in Fig. 2, where the series is represented by SAX word "10 10 11".

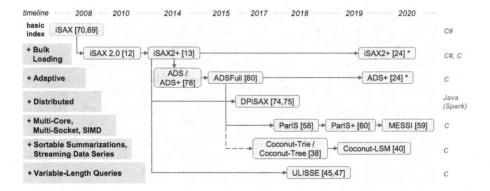

Fig. 1. Lineage of the iSAX family of indexes. Timeline is depicted on the top; implementation languages are marked on the right. Solid arrows denote inheritance of the index design; dashed arrows denote inheritance of some of the design features; the two new versions of iSAX2+ and ADS+ marked with an asterisk support approximate similarity search with deterministic and probabilistic quality guarantees. Source code available by following the links in the corresponding papers.

The SAX representation was later extended to indexable SAX (iSAX) [70], which allows variable cardinality for each character of a SAX representation. An iSAX representation is composed of a set of characters that form a word, and each word represents a data series. In the case of a binary alphabet, with a word size of 3 characters and a maximum cardinality of 2 bits, we could have a set of data series (two in the following example) represented with the following words: $00_2 10_2 01_2$, $00_2 11_2 01_2$, where each character has a full cardinality of 2 bits and each word corresponds to one data series. Reducing the cardinality of the second character in each word, we get for both words the same iSAX representation: $00_2 1_1 01_2$ (1_1 corresponds to both 10 and 11, since the last bit is trailed when the cardinality is reduced). Starting with a cardinality of 1 for each character in the root node and gradually splitting by increasing the cardinality one character at a time, we can build in a top-down fashion the (non-balanced) iSAX tree index [69,70]. These algorithms can be efficiently implemented with bit-wise operations.

The iSAX index supports both approximate and exact similarity search [23]: approximate does not guarantee that it will always find the correct answers (though, in most cases it returns high-quality results [24,70]); exact guarantees that it will always return the correct results. In approximate search, the algorithm uses the iSAX summaries to traverse a single path of the index tree from the root to the most promising leaf, then computes the raw distances between the

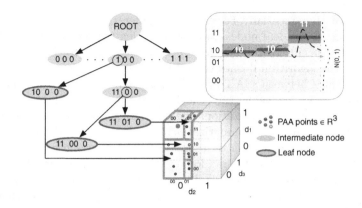

Fig. 2. An example of iSAX and SAX representations [78].

query and each series in the leaf, and return the series with the smallest distance, i.e., the Best-So-Far distance (BSF). Exact search starts with an approximate search that returns a BSF, which is then used to prune the rest of the index leaves; the leaves that cannot be pruned are visited, the raw distances of the series to the query are computed, and the BSF is updated (if needed). At the end of this process, we get the exact answer.

3.2 Bulk-Loading: iSAX 2.0 and iSAX2+

Inserting a large collection of data series into the index iteratively is an expensive operation, involving a high number of disk I/O operations [12,13]. This is because for each time series, we have to store the raw data series on disk, and insert into the index the corresponding iSAX representation. In order to speedup the process of building the index, iSAX 2.0 [12] and iSAX2+ [13] were the first data series indexes (based on the iSAX index) with a bulk loading strategy.

The key idea is to effectively group the data series that will end up in a particular subtree of the index, and process them all together. In order to achieve this goal, we use two main memory buffer layers, namely, the First Buffer Layer (FBL), and the Leaf Buffer Layer (LBL) [13]. The FBL corresponds to the children of the root of the index, while the LBL corresponds to the leaf nodes. The role of the buffers in FBL is to cluster together data series that will end up in the same subtree of the index, rooted in one of the direct children of the root. In contrast, the buffers in LBL are used to gather all the data series of leaf nodes, and flush them to disk.

The algorithm operates in two phases, which alternate until the entire dataset is processed, as follows (for more details, refer to [13]). During Phase 1, the algorithm reads data series and inserts them in the corresponding buffer in the FBL. This phase continues until the main memory is full. Then Phase 2 starts, where the algorithm proceeds by moving the data series contained in each FBL buffer to the appropriate LBL buffers. During this phase, the algorithm processes the buffers in FBL sequentially. For each FBL buffer, the algorithm creates all the necessary internal and leaf nodes, in order to index these data series. When all data series of a specific FBL buffer have been moved down to the corresponding LBL buffers, the algorithm flushes these LBL buffers to disk.

The difference between iSAX 2.0 [12] and iSAX2+ [13] is that the former treats the data series raw values (i.e., the detailed sequence of all the values of the data series) and their summarizations (i.e., the iSAX representations) together, while the latter uses just the summarizations in order to build the index, and only processes the raw values in order to insert them to the correct leaf node. In both cases, the goal is to minimize the random disk I/O, by making sure that the data series that end up in the same leaf node of the index are (temporarily) stored in the same (or contiguous) disk pages. The experiments demonstrate that iSAX 2.0 and iSAX2+ significantly outperform previous approaches, reducing the time required to index 1 billion data series by 72% and 82%, respectively. A recent extension of iSAX2+ supports approximate answers with quality guarantees [24].

3.3 Adaptive Indexing: ADS, ADS+, ADS-Full

Even though iSAX 2.0 and iSAX2+ can effectively cope with very large data series collections, users still have to wait for extended periods of time before the entire index is built and being able to start answering queries.

The Adaptive Data Series (ADS) and ADS+ indexes [78,79] are based on the iSAX 2.0 index, and address the above problem. They perform only a few basic steps, mainly creating the basic skeleton of the index tree, which contains condensed information on the input data series, and are then ready to start answering queries. As queries arrive, ADS fetches data series from the raw data and moves only those data series needed to correctly answer the queries inside the index. Future queries may be completely covered by the contents of the index, or alternatively ADS adaptively and incrementally fetches any missing data series directly from the raw data set. When the workload stabilizes, ADS can quickly serve fully contained queries while as the workload shifts, ADS may temporarily need to perform some extra work to adapt before stabilizing again.

The additional feature of ADS+ (when compared to ADS) is that it does not require a fixed leaf size: it dynamically and adaptively adjusts the leaf size in hot areas of the index. ADS+ uses two different leaf sizes: a big build-time leaf size for optimal index construction, and a small query-time leaf size for optimal access costs. Initially, the index tree is built as in plain ADS, with a constant leaf size, equal to build-time leaf size. In traditional indexes, this leaf size remains the same across the life-time of the index. In our case, when a query that needs to search a partial leaf arrives, ADS+ refines its index structure on-the-fly by recursively splitting the target leaf, until the target sub-leaf becomes smaller or equal to the query-time leaf size.

ADS and ADS+ support the same query answering mechanisms as iSAX2.0 and iSAX2+, but they also introduced the Scan of In-Memory Summarizations (SIMS) algorithm for exact query answering. SIMS starts by an approximate search to compute the BSF, which is then used to compare to the in-memory iSAX summaries of all the series in the collection, and finally, performs a skip-sequential scan of the raw series that were not pruned in the previous step.

Experiments with up to 1 billion data series and 10^5 random approximate queries show that ADS+ answers all queries in less than 5 h, while iSAX 2.0 needs more than 35 h. In turn, ADS+ and iSAX 2.0 are orders of magnitude faster in index creation than KD-Tree [8], R-Tree [29], and X-Tree [9].

In settings where a complete index is required, i.e., when there is a completely random and very large work-load, a full index can also be efficiently constructed using ADS-Full [80]. In the first step, the ADS structure is built by performing a full pass over the raw data file, storing only the iSAX representations at each leaf. In the second step, one more sequential pass over the raw data file is performed, and data series are moved in the correct pages on disk. The benefit of this process is that it completely skips costly split operations on raw data series, leading to a 2x–3x faster creation of the full index, when compared to iSAX 2.0. A recent extension of ADS+ supports approximate answers with quality guarantees [24].

3.4   Parallel and Distributed: ParIS, ParIS+, MESSI, DPiSAX

The continued increase in the rate and volume of data series production with collections that grow to several terabytes in size [53] renders single-core data series indexing technologies inadequate. For example, ADS+ [80], requires >4 min to answer a single exact query on a moderately sized 250 GB sequence collection.

The Parallel Index for Sequences (ParIS) [58], based on ADS+, is the first data series index that takes advantage of modern hardware parallelization, and incorporate the state-of-the-art techniques in sequence indexing, in order to accelerate processing times. ParIS, which is a disk-based index, can effectively operate on multi-core and multi-socket architectures, in order to distribute and execute in parallel the computations needed for both index construction and query answering. Moreover, ParIS exploits the Single Instruction Multiple Data (SIMD) capabilities of modern CPUs, to further parallelize the execution of individual instructions inside each core. Overall, ParIS achieves very good overlap of the CPU computation with the required disk I/O. ParIS+ [60], an alternative of ParIS, completely removes the CPU cost during index creation, resulting in index creation that is purely I/O bounded, and 2.6x faster than ADS+. ParIS+ achieves this by reorganizing the way that the workload is distributed among the worker threads. ParIS and ParIS+ employ the same algorithmic techniques for query answering. The experiments also demonstrate their effectiveness in exact query answering: they are up to 1 order of magnitude faster than ADS+, and up to 3 orders of magnitude faster than the state-of-the-art optimized serial scan method, UCR Suite [63]. We also note that ParIS and ParIS+ have the potential to deliver more benefit as we move to faster storage media.

Still, ParIS+ is designed for disk-resident data and therefore its performance is dominated by the I/O costs it encounters. For instance, ParIS+ answers a 1-NN exact query on a 100 GB dataset in 15 s, which is above the limit for keeping the user's attention (i.e., 10 s), let alone for supporting interactivity in the analysis process (i.e., 100 ms) [26]. The in-MEmory data SerieS Index

(MESSI) [59] is based on ParIS+, and is the first parallel index designed for memory-resident datasets. MESSI effectively uses multi-core and multi-socket architectures in order to concurrently execute the computations needed for both index construction and query answering, and it exploits SIMD. Since MESSI copes with in-memory data series, no CPU cost can be hidden under I/O, and required more careful design choices and coordination of the parallel workers when accessing the required data structures, in order to improve its performance. This led to the development of a more subtle design for the construction of the index and on the development of new algorithms for answering similarity search queries on this index. The results show a further ~4x speedup in index creation time, in comparison to an in-memory version of ParIS+. Furthermore, MESSI answers exact 1-NN queries on 100 GB datasets 6-11x faster than ParIS+, achieving for the first time interactive exact query answering times, at ~50 ms.

In order to exploit parallelism across compute nodes, the Distributed Partitioned iSAX (DPiSAX) [42,74,75] index was developed. DPiSAX is based on iSAX2+, and was designed to operate on top of Spark. DPiSAX uses a sampling phase that allows to balance the partitions of data series across the compute nodes (according to their iSAX representations), which is necessary for efficient query processing. DPiSAX gracefully scales to billions of time series, and a parallel query processing strategy that, given a batch of queries, efficiently exploits the index. The experiments show that DPiSAX can build its index on 4 billion data series in less than 5 h (and one order of magnitude faster than iSAX2+). Also, DPiSAX processes 10 millions 10-NN approximate queries on a 1 billion data series collection in 140 s.

The DPiSAX solution is complementary to the ParIS+ and MESSI solutions, and they could be combined in order to exploit both parallelism and distribution.

3.5   Variable-Length: ULISSE

Despite the fact that data series indexes enable fast similarity search, all existing indexes can only answer queries of a single length (fixed at index construction time), which is a severe limitation. The ULtra compact Index for variable-length Similarity SEarch (ULISSE) [45, 46] is the first, single data series index structure designed for answering similarity search queries of variable length. ULISSE introduces a novel envelope representation that effectively and succinctly summarizes multiple sequences of different lengths. These envelopes are then used to build a tree index that resembles to iSAX2+. ULISSE supports both approximate and exact similarity search, combining disk based index visits with in-memory sequential scans, inspired by ADS+. ULISSE supports non Z-normalized and Z-normalized sequences, and can be used with no changes with both Euclidean Distance and Dynamic Time Warping, for answering k-NN and ϵ-range queries [47].

The experimental results show that ULISSE is several times, and up to orders of magnitude more efficient in terms of both space and time cost, when compared to competing approaches (i.e., UCR Suite, MASS, and CMRI) [45,47].

3.6   Sortable Summarizations: Coconut-Trie/Tree/LSM

We observe that a shortcoming of the indexes presented earlier is that their design is based on summarizations [14,44] (used as keys by the index) that are unsortable. Thus, sorting based on these summarizations would place together data series that are similar in terms of their beginning, i.e., the first segment, yet arbitrarily far in terms of the rest of the segments. Hence, existing summarizations cannot be sorted while keeping similar data series next to each other in the sorted order. This leads to top-down index building (resulting in many small random disk I/Os and non-contiguous nodes), and prefix-based node-splitting (resulting in low fill-factors for leaf nodes), which negatively affect time performance and disk space occupancy.

The **Co**mpact and **Con**tiguous Sequence Infrastructure (Coconut) index [38,39] was developed in order to address these problems, by transforming the iSAX summarization into a *sortable summarization*. The core idea is interweaving the bits that represent the different segments, such that the more significant bits across all segments precede all less significant bits. As a result, Coconut is the first technique for sorting data series based on their summarizations that can lead to bottom-up creation of balanced indexes: the series are positioned on a z-order curve [52], in a way that similar data series are close to each other. Indexing based on sortable summarizations has the same ability as existing summarizations to prune the search space. Coconut supports bulk-loading techniques and log-structured updates to enable maintaining a contiguous index. This eliminates random I/O during construction, updating and querying. Furthermore, Coconut is able to split data series across nodes by sorting them and using the median value as a splitting point, leading to data series being packed more densely into leaf nodes (i.e., at least half full). We studied Coconut-Trie and Coconut-Tree, which split data series across nodes based on common prefixes and median values, respectively. Coconut-Trie, which is similar to an ADS+ index in structure, dominates the state-of-the-art in terms of query speed because it creates contiguous leaves. Coconut-Tree, based on a B+-Tree index, dominates Coconut-Trie and the state-of-the-art in terms of index construction speed, query (using SIMS) speed and storage overheads because it creates a contiguous, balanced index that is also densely populated. Finally, Coconut-LSM [39,40], that is based on an LSM tree index, supports efficient log-structured updates and variable-size window queries over different windows of the data based on recency.

Overall, across a wide range of workloads and datasets, Coconut-Tree improves both construction speed and storage overheads by one order of magnitude and query speed by two orders of magnitude relative to DSTree and ADS. Coconut-LSM supports updates without degrading query throughput, and is able to narrow the search scope temporally. This improves query throughput by a further 2–3 orders of magnitudes in our experiments for queries over recent data, thus, making Coconut-LSM an efficient solution for streaming data series.

4   Discussion and Open Research Directions

Despite the strong increasing interest in data series management systems [83], existing approaches (e.g., based on DBMSs, Column Stores, TSMSs, or Array Databases) do not provide a viable solution, since they have not been designed for managing and processing sequence data as first class citizens: they do not offer a suitable storage model, declarative query language, or optimization mechanism. Moreover, they lack auxiliary data structures (such as indexes), that can support a variety of sequence query workloads in an efficient manner. For example, they do not have native support for similarity search [31,53], and therefore, cannot efficiently support complex analytics on very large data series collections.

Current solutions for processing data series collections, in various domains, are mostly ad hoc (and hardly scalable), requiring huge investments in time and effort, and duplication of effort across different teams. For this reason new data management technologies should be developed; albeit ones that will meet their requirements for processing and analyzing very large sequence collections.

An interesting and challenging research direction is to design and develop a general purpose Sequence Management System, able to cope with big data series (very large and continuously growing collections of data series with diverse characteristics, which may have uncertainty in their values), by transparently optimizing query execution, and taking advantage of new management and query answering techniques, as well as modern hardware [53,55]. Just like databases abstracted the relational data management problem and offered a black box solution that is now omnipresent, the proposed system will enable users and analysts that are not experts in data series management to tap in the goldmine of the massive and ever-growing data series collections they (already) have.

Our preliminary results, including the first data series similarity search benchmark [81,82], and indexing algorithms that can be efficiently bulk-loaded [12,13,38–40], adapt to the query workload [78–80], support similarity queries of varying length [45,46,48,49], take into account uncertainty [19,20], and exploit multi-cores [58–60] and distributed platforms (e.g., Apache Spark) [42,74,75], are promising first steps. Nevertheless, much progress is still needed along the directions mentioned above. This is especially true for query optimization, since earlier work has shown that different techniques and algorithms perform better for different query workloads and data and hardware characteristics [23]. Trying to further optimize query execution times, techniques that provide approximate answers, and in particular answers with (deterministic, or probabilistic) guarantees on the associated error bounds [23,24], can be very useful. The same is true for techniques that provide progressive answers [28], which can also lead to significant speedup, while guaranteeing the desired levels of accuracy.

It would also be interesting to develop an index that combines all (or most of) the features mentioned earlier, namely, support for progressive exact and approximate queries of variable length, running on modern hardware in parallel and distributed environments. Given the way that these index solutions have been developed, i.e., by building on top of one another, combining the various features in a single solution seems feasible.

Note that, even though the indexes we presented have been developed for data series, they are equally applicable to and extremely efficient in the case of general high-dimensional vectors [23,24]. This opens up several exciting application opportunities, including in deep learning analysis pipelines, where we often need to perform similarity search in high-dimensional vector embeddings.

5 Conclusions

In this work, we discussed the evolution of the iSAX family of indexes, which represent the current state-of-the-art in several variations of the problem of indexing for similarity search in very large data series collections. We reviewed the basic design decisions behind these indexes, and contrasted their strong points. The presentation (for the first time together) of all these indexes contributes to the better understanding of which particular problem each one solves, how their features could be combined, and what the opportunities for future work are.

Acknowledgements. I would like to thank my collaborators (in alphabetical order): R. Akbarinia, H. Benbrahim, A. Bezerianos, A. Camerra, M. Dallachiesa, N. Dayan, K. Echihabi, A. Gogolou, P. Fatourou, J. Gehrke, S. Idreos, I. Ilyas, E. Keogh, B. Kolev, H. Kondylakis, O. Levchenko, M. Linardi, Y. Lou, F. Masseglia, K. Mirylenka, B. Nushi, B. Peng, T. Rakthanmanon, D. Shasha, J. Shieh, T. Tsandilas, P. Valduriez, and D.-E. Yagoubi. Special thanks go to K. Zoumpatianos.

References

1. ADHD-200 (2011). http://fcon_1000.projects.nitrc.org/indi/adhd200/
2. Sloan Digital Sky Survey (2015). https://www.sdss3.org/dr10/data_access/volume.php
3. Agrawal, R., Faloutsos, C., Swami, A.: Efficient similarity search in sequence databases. In: Lomet, D.B. (ed.) FODO 1993. LNCS, vol. 730, pp. 69–84. Springer, Heidelberg (1993). https://doi.org/10.1007/3-540-57301-1_5
4. An, N., Kothuri, R.K.V., Ravada, S.: Improving performance with bulk-inserts in Oracle R-trees. In: VLDB, pp. 948–951. VLDB Endowment (2003)
5. Assent, I., Krieger, R., Afschari, F., Seidl, T.: The TS-tree: efficient time series search and retrieval. In: EDBT (2008)
6. Aßfalg, J., Kriegel, H.-P., Kröger, P., Kunath, P., Pryakhin, A., Renz, M.: Similarity search on time series based on threshold queries. In: Ioannidis, Y., et al. (eds.) EDBT 2006. LNCS, vol. 3896, pp. 276–294. Springer, Heidelberg (2006). https://doi.org/10.1007/11687238_19
7. Bagnall, A.J., Cole, R.L., Palpanas, T., Zoumpatianos, K.: Data series management (Dagstuhl seminar 19282). Dagstuhl Rep. **9**(7), 24–39 (2019)
8. Bentley, J.L.: Multidimensional binary search trees used for associative searching. Commun. ACM **18**(9), 509–517 (1975)
9. Berchtold, S., Keim, D.A., Kriegel, H.-P.: The X-tree: an index structure for high-dimensional data. In: VLDB, pp. 28–39 (1996)
10. Berndt, D.J, Clifford, J.: Using dynamic time warping to find patterns in time series. In: AAAIWS, pp. 359–370 (1994)

11. Bu, Y., Leung, T.W., Fu, A.W.C., Keogh, E., Pei, J., Meshkin, S.: WAT: finding top-k discords in time series database. In: SDM, pp. 449–454 (2007)
12. Camerra, A., Palpanas, T., Shieh, J., Keogh, E.: iSAX 2.0: indexing and mining one billion time series. In: ICDM (2010)
13. Camerra, A., Shieh, J., Palpanas, T., Rakthanmanon, T., Keogh, E.J.: Beyond one billion time series: indexing and mining very large time series collections with iSAX2+. KAIS **39**(1), 123–151 (2014). https://doi.org/10.1007/s10115-012-0606-6
14. Chakrabarti, K., Keogh, E., Mehrotra, S.: Locally adaptive dimensionality reduction for indexing large time series databases. ACM Trans. Database Syst. (TODS) **27**(2), 188–228 (2002)
15. Chakrabarti, K., Keogh, E., Mehrotra, S., Pazzani, M.: Locally adaptive dimensionality reduction for indexing large time series databases. In: SIGMOD (2002)
16. Chan, K.-P., Fu, A.-C.: Efficient time series matching by wavelets. In: ICDE (1999)
17. Chandola, V., Banerjee, A., Kumar, V.: Anomaly detection: a survey. ACM Comput. Surv. **41**(3), 1–58 (2009)
18. Chen, Y., Nascimento, M.A., Ooi, B.C., Tung, A.K.H.: SpADe: on shape-based pattern detection in streaming time series. In: ICDE (2007)
19. Dallachiesa, M., Nushi, B., Mirylenka, K., Palpanas, T.: Uncertain time-series similarity: return to the basics. PVLDB **5**(11), 1662–1673 (2012)
20. Dallachiesa, M., Palpanas, T., Ilyas, I.F.: Top-k nearest neighbor search in uncertain data series. PVLDB **8**(1), 13–24 (2014)
21. Das, G., Gunopulos, D., Mannila, H.: Finding similar time series. In: Komorowski, J., Zytkow, J. (eds.) PKDD 1997. LNCS, vol. 1263, pp. 88–100. Springer, Heidelberg (1997). https://doi.org/10.1007/3-540-63223-9_109
22. Ding, H., Trajcevski, G., Scheuermann, P., Wang, X., Keogh, E.: Querying and mining of time series data: experimental comparison of representations and distance measures. In: PVLDB (2008)
23. Echihabi, K., Zoumpatianos, K., Palpanas, T., Benbrahim, H.: The Lernaean Hydra of data series similarity search: an experimental evaluation of the state of the art. PVLDB **12**(2), 112–127 (2018)
24. Echihabi, K., Zoumpatianos, K., Palpanas, T., Benbrahim, H.: Return of the Lernaean Hydra: experimental evaluation of data series approximate similarity search. PVLDB **13**, 403–420 (2019)
25. Soisalon-Soininen, E., Widmayer, P.: Single and bulk updates in stratified trees: an amortized andworst-case analysis. In: Klein, R., Six, H.-W., Wegner, L. (eds.) Computer Science in Perspective. LNCS, vol. 2598, pp. 278–292. Springer, Heidelberg (2003). https://doi.org/10.1007/3-540-36477-3_21
26. Fekete, J.-D., Primet, R.: Progressive analytics: a computation paradigm for exploratory data analysis. CoRR (2016)
27. Gogolou, A., Tsandilas, T., Palpanas, T., Bezerianos, A.: Comparing similarity perception in time series visualizations. IEEE TVCS **25**(1), 523–533 (2019)
28. Gogolou, A., Tsandilas, T., Palpanas, T., Bezerianos, A.: Progressive similarity search on time series data. In: Workshops of the EDBT/ICDT (2019)
29. Guttman, A.: R-trees: a dynamic index structure for spatial searching. In: SIGMOD (1984)
30. Huijse, P., Estévez, P.A., Protopapas, P., Principe, J.C., Zegers, P.: Computational intelligence challenges and applications on large-scale astronomical time series databases. IEEE Comput. Int. Mag. **9**(3), 27–39 (2014)
31. Jensen, S.K., Pedersen, T.B., Thomsen, C.: Time series management systems: a survey. IEEE Trans. Knowl. Data Eng. **29**(11), 2581–2600 (2017)

32. Seeger, B., Van den Bercken, J.: An evaluation of generic bulk loading techniques. In: VLDB, pp. 461–470 (2001)
33. Widmayer, P., Van den Bercken, J., Seeger, B.: A generic approach to bulk loading multidimensional index structures. In: VLDB (1997)
34. Kashino, K., Smith, G., Murase, H.: Time-series active search for quick retrieval of audio and video. In: ICASSP (1999)
35. Kashyap, S., Karras, P.: Scalable KNN search on vertically stored time series. In: KDD (2011)
36. Keogh, E., Chakrabarti, K., Pazzani, M., Mehrotra, S.: Dimensionality reduction for fast similarity search in large time series databases. KAIS 3(3), 263–286 (2000). https://doi.org/10.1007/PL00011669
37. Keogh, E.J., Palpanas, T., Zordan, V.B., Gunopulos, D., Cardle, M.: Indexing large human-motion databases. In: VLDB, pp. 780–791 (2004)
38. Kondylakis, H., Dayan, N., Zoumpatianos, K., Palpanas, T.: Coconut: a scalable bottom-up approach for building data series indexes. In: PVLDB (2018)
39. Kondylakis, H., Dayan, N., Zoumpatianos, K., Palpanas, T.: Coconut palm: static and streaming data series exploration now in your palm. In: SIGMOD, pp. 1941–1944 (2019)
40. Kondylakis, H., Dayan, N., Zoumpatianos, K., Palpanas, T.: Coconut: sortable summarizations for scalable indexes over static and streaming data series. VLDBJ $\mathbf{28}$, 847–869 (2019). https://doi.org/10.1007/s00778-019-00573-w
41. Arge, L., Hinrichs, K., Vahrenhold, J., et al.: Efficient bulk operations on dynamic R-trees. Algorithmica $\mathbf{33}$(1), 104–128 (2002). https://doi.org/10.1007/s00453-001-0107-6
42. Levchenko, O., et al.: Distributed algorithms to find similar time series. In: ECML/PKDD (2019)
43. Li, C.-S., Yu, P., Castelli, V.: HierarchyScan: a hierarchical similarity search algorithm for databases of long sequences. In: ICDE (1996)
44. Lin, J., Keogh, E., Lonardi, S., Chiu, B.: A symbolic representation of time series, with implications for streaming algorithms. In: DMKD (2003)
45. Linardi, M., Palpanas, T.: Scalable, variable-length similarity search in data series: the ULISSE approach. PVLDB $\mathbf{11}$(13), 2236–2248 (2018)
46. Linardi, M., Palpanas, T.: ULISSE: ULtra compact index for variable-length similarity SEarch in data series. In: ICDE (2018)
47. Linardi, M., Palpanas, T.: Scalable data series subsequence matching with ULISSE. Technical Report (2020)
48. Linardi, M., Zhu, Y., Palpanas, T., Keogh, E.J.: Matrix profile X: VALMOD - scalable discovery of variable-length motifs in data series (2018)
49. Linardi, M., Zhu, Y., Palpanas, T., Keogh, E.J.: VALMOD: a suite for easy and exact detection of variable length motifs in data series. In: SIGMOD (2018)
50. Mirylenka, K., Dallachiesa, M., Palpanas, T.: Correlation-aware distance measures for data series. In: EDBT, pp. 502–505 (2017)
51. Mirylenka, K., Dallachiesa, M., Palpanas, T.: Data series similarity using correlation-aware measures. In: SSDBM (2017)
52. Morton, G.M.: A Computer Oriented Geodetic Data Base and a New Technique in File Sequencing. International Business Machines Company, Ottawa (1966)
53. Palpanas, T.: Data series management: the road to big sequence analytics. SIGMOD Rec. $\mathbf{44}$, 47–52 (2015)

54. Palpanas, T.: Big sequence management: a glimpse of the past, the present, and the future. In: Freivalds, R.M., Engels, G., Catania, B. (eds.) SOFSEM 2016. LNCS, vol. 9587, pp. 63–80. Springer, Heidelberg (2016). https://doi.org/10.1007/978-3-662-49192-8_6

55. Palpanas, T.: The parallel and distributed future of data series mining. In: High Performance Computing & Simulation (HPCS) (2017)

56. Palpanas, T., Beckmann, V.: Report on the first and second interdisciplinary time series analysis workshop (ITISA). ACM SIGMOD Rec. **48**(3), 36–40 (2019)

57. Palpanas, T., Vlachos, M., Keogh, E.J., Gunopulos, D.: Streaming time series summarization using user-defined amnesic functions. IEEE Trans. Knowl. Data Eng. **20**(7), 992–1006 (2008)

58. Peng, B., Fatourou, P., Palpanas, T.: Paris: the next destination for fast data series indexing and query answering. In: IEEE BigData, pp. 791–800 (2018)

59. Peng, B., Fatourou, P., Palpanas, T.: MESSI: in-memory data series indexing. In: ICDE (2020)

60. Peng, B., Fatourou, P., Palpanas, T.: Paris+: data series indexing on multi-core architectures. In: TKDE (2020)

61. Rafiei, D.: On similarity-based queries for time series data. In: ICDE (1999)

62. Rafiei, D., Mendelzon, A.: Similarity-based queries for time series data. In: SIGMOD (1997)

63. Rakthanmanon, T.: Searching and mining trillions of time series subsequences under dynamic time warping. In: KDD (2012)

64. Raza, U., Camerra, A., Murphy, A.L., Palpanas, T., Picco, G.P.: Practical data prediction for real-world wireless sensor networks. TKDE **27**(8), 2231–2244 (2015)

65. Choubey, R., Chen, L., Rundensteiner, E.A.: GBI: a generalized R-tree bulk-insertion strategy. In: Güting, R.H., Papadias, D., Lochovsky, F. (eds.) SSD 1999. LNCS, vol. 1651, pp. 91–108. Springer, Heidelberg (1999). https://doi.org/10.1007/3-540-48482-5_8

66. Schäfer, P., Högqvist, M.: SFA: a symbolic fourier approximation and index for similarity search in high dimensional datasets. In: EDBT (2012)

67. Shasha, D.: Tuning time series queries in finance: case studies and recommendations. IEEE Data Eng. Bull. **22**(2), 40–46 (1999)

68. Shieh, J., Keogh, E.: iSAX: indexing and mining terabyte sized time series. In: SIGKDD, pp. 623–631 (2008)

69. Shieh, J., Keogh, E.: iSAX: disk-aware mining and indexing of massive time series datasets. DMKD **19**(1), 24–57 (2009). https://doi.org/10.1007/s10618-009-0125-6

70. Shieh, J., Keogh, E.J.: iSAX: indexing and mining terabyte sized time series. In: KDD, pp. 623–631 (2008)

71. Wang, X., Mueen, A., Ding, H., Trajcevski, G., Scheuermann, P., Keogh, E.: Experimental comparison of representation methods and distance measures for time series data. Data Min. Knowl. Discov. **26**(2), 275–309 (2013)

72. Wang, Y., Wang, P., Pei, J., Wang, W., Huang, S.: A data-adaptive and dynamic segmentation index for whole matching on time series. PVLDB **6**(10), 793–804 (2013)

73. Liao, T.W.: Clustering of time series data - a survey. Pattern Recogn. **38**(11), 1857–1874 (2005)

74. Yagoubi, D.-E., Akbarinia, R., Masseglia, F., Palpanas, T.: DPiSAX: massively distributed partitioned iSAX. In: ICDM (2017)

75. Yagoubi, D.-E., Akbarinia, R., Masseglia, F., Palpanas, T.: Massively distributed time series indexing and querying. TKDE **32**(1), 108–120 (2020)

76. Ye, L., Keogh, E.J.: Time series shapelets: a new primitive for data mining. In: KDD (2009)
77. Yi, B., Faloutsos, C.: Fast time sequence indexing for arbitrary Lp norms. In: VLDB (2000)
78. Zoumpatianos, K., Idreos, S., Palpanas, T.: Indexing for interactive exploration of big data series. In: SIGMOD (2014)
79. Zoumpatianos, K., Idreos, S., Palpanas, T.: RINSE: interactive data series exploration with ADS+. PVLDB 8(12), 1912–1923 (2015)
80. Zoumpatianos, K., Idreos, S., Palpanas, T.: ADS: the adaptive data series index. VLDB J. 25, 843–866 (2016). https://doi.org/10.1007/s00778-016-0442-5
81. Zoumpatianos, K., Lou, Y., Ileana, I., Palpanas, T., Gehrke, J.: Generating data series query workloads. VLDB J. 27(6), 823–846 (2018). https://doi.org/10.1007/s00778-018-0513-x
82. Zoumpatianos, K., Lou, Y., Palpanas, T., Gehrke, J.: Query workloads for data series indexes. In: KDD (2015)
83. Zoumpatianos, K., Palpanas, T.: Data series management: fulfilling the need for big sequence analytics. In: ICDE (2018)

Data Integration

Proximity-Based Federation of Smart Objects: Its Application Framework for Complex Secure Federation Scenarios

Yuzuru Tanaka[1,2,3]([✉])

[1] Hokkaido University, Sapporo, Japan
`tanaka.yzr@ist.hokudai.ac.jp`
[2] Comprehensive Research Organization for Science and Society (CROSS), Tsukuba, Japan
[3] Department of Computing Science, University of Alberta, Edmonton, AB, Canada

Abstract. This paper focuses first on the formal modeling of complex application scenarios using autonomic proximity-based federation among smart objects with wireless network connectivity, and then on a new framework for complex secure federation scenarios. Our modeling consists of three different levels. In the first-level modeling, each smart object is modeled as a set of ports, each of which represents an I/O interface for a function of this smart object to interoperate with some function of another smart object. The federation between a pair of smart objects having a pair of ports of the same type with opposite polarities is modeled as the port matching between these two ports. The second-level modeling describes the dynamic change of the federation structure among smart objects as a graph rewriting system, where each node and each directed link respectively represent a smart object and a connection between two smart objects. The third-level modeling uses a binary autocatalytic-reaction network to describe each complex federation scenario in which more than one federation are involved, and an output federation of a reaction may work either as an input federation of another reaction and/or a catalyst to activate another composition or decomposition reaction. Based on these models previously proposed by the current author, this paper proposes a new simplified application framework for implementing any complex application scenario describable as a binary autocatalytic-reaction network as a graph rewriting system of smart objects, and then proposes a new framework-level solution to the secure federation of smart objects, which is independent from the encryption-based technologies for secure communication between two smart objects.

Keywords: Smart object · Proximity-based federation · IoT · Pervasive computing · Ubiquitous computing · Graph rewriting system · Binary autocatalytic-reaction network

1 Introduction

In the age of smart phones, IC cards, and IoT, we are surrounded by a huge number of smart objects, i.e., intelligent devices with wireless communication capabilities ranging from P2P (peer-to-peer) to cellphone communications. Some of them are wearable or in-vehicle mobile ones, while the others are stationary ones. However, it is often pointed out

© Springer Nature Switzerland AG 2020
G. Flouris et al. (Eds.): ISIP 2019, CCIS 1197, pp. 87–100, 2020.
https://doi.org/10.1007/978-3-030-44900-1_6

both by theoreticians and by practitioners that the lack of a formal computation model and an application framework capable of context modeling and complex application-scenario description to cover the application diversity of smart objects and their federations is the main reason why most existing applications still remain within the scope of three stereotyped scenarios [1–3], i.e., (1) the location transparent service continuation, (2) the location-, and/or situation-aware service provision, and (3) the dynamic federation among smart objects through the Internet, i.e., their web-based federation. The first one focuses on the ubiquity of services, while the second focuses on the context-dependent services. The third one is called IoT.

In his previous papers [3–5], the current author proposed three different levels of formal modeling for describing complex application scenarios using more than one dynamic proximity-based federation reactions of smart objects, where "smart objects" denote computing devices with wireless communication capabilities. "Proximity-based federation" denotes federation that is autonomously activated by the proximity among smart objects, while "federation" denotes dynamic composition of smart objects for their interoperation, and is equivalent to Bill Joy's concept of service federation between service-requesting and service-providing smart objects through wireless connection [6].

The first level modeling formally defines a smart object and describes the federation between one smart object and another within the scope of the former as the port matching process. The "scope" of a smart object denotes the set of all the smart objects within its current wireless communication range, i.e., all the accessible smart objects, or strictly speaking, all the identifiable smart objects. A smart object may not be able to access another smart object, even if it identifies the other, because of the access request denial. The second level describes the dynamic change of federation structures as a graph rewriting system with each node representing a smart object and each directed edge representing a channel connection between a pair of smart objects. The third level deals with complex application scenarios, in each of which more than one smart object federation are involved. It describes each complex application scenario as a binary autocatalytic-reaction network. It consists of binary (composition/decomposition) catalytic reactions each of which does not deal with the quantity of each of its input, output, nor catalyst compounds, but only with the existence or absence of each of them. Each composition reaction represents the federation of given input federations to compose an output federation with or without the help of a catalyst federation. Each decomposition reaction represents the defederation of an input federation into output federations with the help of a catalyst federation. We do not consider any decomposition reaction without the help of a catalyst federation, since it implies that the input federation is too unstable to exist. A binary autocatalytic-reaction network is a set of composition and/or decomposition reactions in which the output of some reaction may work as an input and/or a catalyst of another reaction.

Based on these three levels of formal modeling, the present author proposed a novel middleware framework for the rapid development of complex application scenarios using the proximity-based federation of smart objects [3, 5]. His framework uses a special type of smart objects working as a tag for identifying each different type of smart objects. These tag objects are called nucleotide smart objects since they work as nucleotides in the biological RNA (ribonucleic acid) replication process. For each different tag, we can use the same prototype tag-smart-object that allows us to manually set its type. His framework defined the generic whole set of rewriting rules for this prototype tag-smart-object to execute.

His previous papers mainly focused on binary catalytic reactions using immobile catalysts as contexts and showed only one simple example of binary autocatalytic-reaction networks without using any context. Here, in this paper, we will focus only on complex binary autocatalytic reaction networks without using any context. Instead of extracting the generic federation mechanism from each smart object, encapsulating it into nucleotide smart objects working as tagging objects, and formalizing the federation reaction mechanism in a generic way as the bio-inspired RNA replication mechanism, the new approach directly implements the federation-reaction mechanism in each involving smart object as its capability, which can be simply defined as its graph rewriting rules. This enables us to define the federation capability of each smart object much more easily for an arbitrarily given complex application scenario.

In addition, this paper newly proposes a framework-level solution to the secure federation of smart objects, which is independent from the encryption-based technologies for secure communication between two smart objects and can be used together with them to increase the federation security.

2 Formal Modeling of Smart Objects

2.1 Smart Objects and Their Port Matching [3, 5]

Each smart object communicates with another one through a peer-to-peer communication facility, which is either a direct cable connection or a wireless connection. Some smart objects may have WiFi communication and/or cellular phone communication facilities for their Internet connection. These different types of wireless connections are all proximity-based connections, i.e., each of them has a distance range of wireless communication. We model this by a function $scope(o)$, which denotes a set of smart objects that are currently accessible, or more strictly speaking, identifiable by a smart object o.

Each smart object is modeled as a set of ports. Each port consists of a port type and its polarity, i.e., either a positive polarity '+' or a negative polarity '−'. A smart object that provides a service of type 'stype' has a service-providing port +stype. A smart object has a service-requesting port −stype if it requests a service of type 'stype'. A service type 'stype' may or may not depend on its providing smart object or its type. Sometimes, we may consider the object identifier or the object type of a smart object as a service type. A service-providing port +oid of a smart object with 'oid' as its object identifier denotes that this object publicizes its object identifier so that any other object that knows this object identifier and exists within its proximity $scope(oid)$ can access this object through its service requesting port −oid. Similarly, A service-providing port +otype of a smart object o with 'otype' as its object type denotes that this object publicizes its object type 'otype' so that any other object that knows this object type 'otype' and exists within its proximity $scope(o)$ can access this object through its service requesting port −otype.

An object o_1 with a service-requesting port −stype can access a service 'stype' provided by another object o_2 in $scope(o_1)$ as follows. The object o_1 first internally sends a message to its port $-$stype, which delegates this message to the $+$stype port of o_2 through wireless communication. This service-providing port $+$stype of o_2 then invokes the service of type 'stype' defined in o_2 with this message to receive a return value from it. Then the service-providing port $+$stype of o_2 returns this value to the port $-$stype of o_1. Finally, the process which initially sent a massage to this $-$stype port receives this return value from the service.

Federation of a smart object o with another smart object o' in its scope *scope*(o) is initiated by a program running on o or on some other activating smart object that can access both of these two objects. The activating object here denotes the smart object that executes this program. This program detects either a specific user operation on o or on the activating object, a change of *scope*(o), or some other event on o or on the activating object as a trigger to initiate federation. The initiation of federation with o' by a smart object o or by some other activating object first checks if o' exists in *scope*(o), and, if yes, it performs the port matching between the ports of o and the ports of o'. As the result, every port $-$p in o is connected with a port $+$p in o' by a channel identified by their shared port type p. We assume that ports are not internally matched with each other to set any channel within a single object.

The same smart object may be involved in more than one different channel. The maximum number of channels in which the same port can be involved is called the arity of this port. In our modeling, we assume that each service-requesting port and each service-providing port may have arbitrary arities unless otherwise specified. In order to specify that a port $\pm$ p has the arity n, we use the notation $\pm$ p(n).

2.2 Graph Rewriting System

The second-level formal modeling focuses on the dynamic change of federation structures among smart objects. It describes a system of smart objects as a directed graph in which each node represents either a smart object or a port, and each directed edge represents either a channel or a proximity relationship. A smart object node and a port node are respectively represented by a bigger (white or gray) circle and a smaller black circle. A channel and a proximity relationship are represented respectively by a black arrow and a gray arrow. Each gray arrow denotes that the pointed smart object is in the scope of the pointing smart object. A port node with an outgoing (or incoming) channel edge p to (from) an object node o denotes that this object o has a service-providing (service-requesting) port $+$p ($-$p), and that it is not involved in any federation yet. Each object node has its state and its type. Smart objects of the same type share the same port set and the same functions. The formalization with graph rewriting rules aims to describe the dynamic change of the channel connections among smart objects through the activation of federation rules, and to hide all the details about the execution of service functions.

Each rewriting rule is specified as a combination of the following four types of rules, i.e., port activation/deactivation rules, state setting rules, channeling rules, and path-dependency rules [3, 5]. In this paper, we only need to use channeling rules for the description of our basic framework, and path-dependency rules in Sect. 4 for the extension of our framework for secure federation. In each of the following rules, there always exists only one gray smart object node. This gray smart object node called the rule-activation node indicates that this rule is stored in this smart object node and executed by this node. Each type of rules is carefully designed to satisfy reasonable hardware and performance constraints of the smart objects of our concern so that it can be executed locally without assuming the accessibility to any global information about the current overall federation structure. The left-hand side of each rule specifies the condition for this rule to be executed. The rule-activation node should be able to check all the specification conditions in this condition part such as those on smart object states, smart object types, port types, port availabilities, channel types, channel connections, and proximity relations of this activation node itself and of all the other nodes specified in the condition part. This implies that the rule-activation node should be able to access all these nodes through channels. The right-hand side of each rule specifies the actions to be executed by the rule-activation node, which should be able to perform these actions directly or to instruct other nodes through channels to perform these actions. Each of the rules defined in each smart object is periodically checked in their definition order if its condition part holds true. Its action part is immediately executed if its condition is satisfied. Otherwise its action is neglected. Then the next rule in the definition order is immediately checked.

For example, Fig. 1 shows the form of channeling rules for setting channels. In each rule in Fig. 1 (a), the rule-activation smart object node (i.e., the gray node) can activate or deactivate a specified port of the left smart object node through the channel path σ (, i.e., a sequence of consecutive channels in the same direction,) to establish or to break the corresponding channel to its neighboring smart object node that is pointed to by a gray arrow. The length of σ may be zero. The smart object that can be reached from a smart object o by a channel path σ is called the σ object of o and denoted by σ(o).

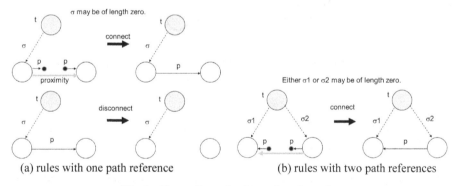

(a) rules with one path reference   (b) rules with two path references

Fig. 1. Channeling rules for setting channels.

In Fig. 1 (b), the rule-activation smart object uses −p and +p ports to establish a channel of type p between the two smart objects. The length of either σ1 or σ2 may be zero.

2.3 Smart Objects and Their Port Matching [3, 5]

A linear federation $o_1 o_2 ... o_n$ of n smart objects o_1, o_2, ..., o_n denotes a sequence of smart objects in which, for each $i = 1$, ..., n-1, there is an L channel from o_{i+1} to o_i. For two linear federations or smart objects X and Y, their federation XY denotes a liner federation in which the first smart object in Y spans an L channel to the last smart object in X. The type of a linear federation $o_1 o_2 ... o_n$ is defined as the concatenation of the object types of o_1, o_2, ..., o_n.

Our third level modeling uses a binary catalytic composition reaction and a binary catalytic decomposition reaction to respectively represent federation and defederation of linear federations of smart objects [3, 5]. Figure 2 lists up all kinds of composition reactions and decomposition reactions, where X, Y, and C denote smart object types or linear federation types, and XY denotes the type of a linear federation of two linear federations of types X and Y. Any decomposition reaction without any catalyst means autonomous decomposition, which indicates that its input linear federation is unstable. Therefore, we do not consider such decomposition reactions, which is shown in gray in Fig. 2, in our binary autocatalytic-reaction network modeling.

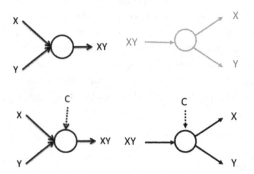

Fig. 2. Composition and decomposition reactions with and without catalysts.

3 Implementing Any Binary Autocatalytic-Reaction Network with Graph Rewriting Rules

A binary catalytic composition reaction with X, Y, and C as two input federations and one catalyst federation to compose XY can be implemented by additionally defining the graph rewriting rules shown in Fig. 3 in this order in the smart object type C_n, i.e., the type of the rightmost smart object of C. The channel P* denotes a temporary channel from $-P$ port to $+P$ port. The channel P* is used for the rightmost smart object C_n of the catalyst C to check the condition of Rule 2. If the condition of Rule 2 does not hold, then this temporary channel is immediately broken by the C_n type smart object using Rule 3. Once the rightmost smart object of type X_h in X and the rightmost smart object of type Y_k in Y are both linked from the rightmost smart object of type C_n in the catalyst C with two P channels, Rule 4 is immediately applied for the smart object of type C_n to span an L channel from the leftmost smart object of Y to the rightmost smart object of X for composing a federation XY. Then the condition part of Rule 5 becomes satisfied, and the execution of this rule by the smart object of type C_n immediately breaks the two P channels between the catalyst C and the federation XY to separate the federation XY from C.

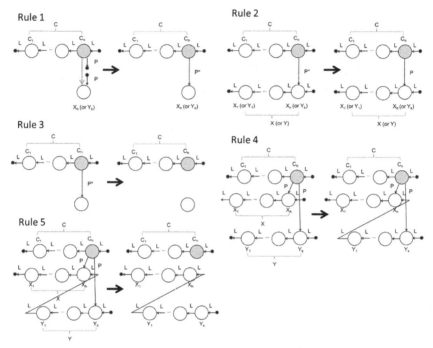

Fig. 3. Graph rewriting rules for a composition of XY from two federations X and Y with the help of a catalyst federation C.

A composition reaction without using any catalyst to compose a federation XY from two input federations X and Y can be implemented by additionally defining the graph rewriting rules shown in Fig. 4 in this order in the smart object type Y_k, i.e., the type of the rightmost smart object of Y. The role of the temporary channel P* is the same as in Fig. 3. Once the rightmost smart object of type X_h in X is linked from the rightmost smart object of type Y_k in Y through a P channel, Rule 4 in Fig. 4 is immediately executed by the smart object of type Y_k to break the P channel and to span an L channel from the leftmost smart object of Y to the rightmost smart object of X for composing a federation XY.

A catalytic decomposition with XY and C as the input and the catalyst to produce X and Y can be implemented by additionally defining the graph rewriting rules shown in Fig. 5 in this order in the smart object type C_n, i.e., the type of the rightmost smart object of C.

Using the above-mentioned general rules to implement three different types of catalytic reactions, any complex application scenario defined as a binary autocatalytic-reaction network can be easily implemented. For example, an application scenario in

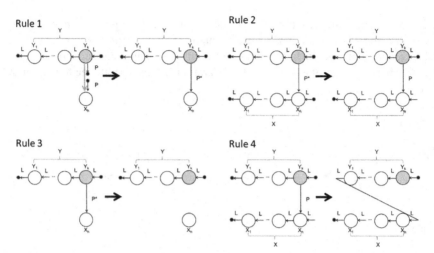

Fig. 4. Graph rewriting rules for a composition of XY from two federations X and Y without the help of any catalyst.

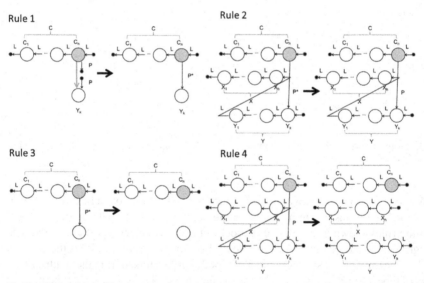

Fig. 5. Graph rewriting rules for a decomposition of XY to two federations X and Y with the help of a catalyst federation C.

Fig. 6 can be implemented by additionally defining rewriting rules in smart object types B and D as shown in Fig. 7.

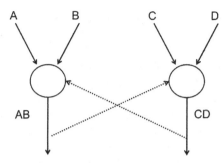

Fig. 6. An example binary autocatalytic-reaction network.

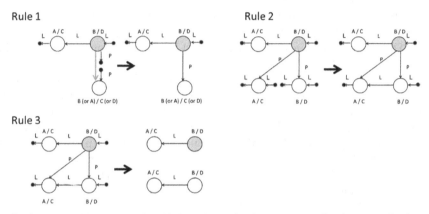

Fig. 7. Graph rewriting rules defined in B and D to implement the application scenario shown in Fig. 6.

4   A Framework-Level Solution to the Secure Federation

While one typical way of increasing the security of federation is the encryption of messages that are exchanged between ports of different smart objects, we will consider in this section a framework-level solution as its complementary way to increase the federation security. A basic idea is the encapsulation of federation so that no one may be able to see nor to interact with any protected ports of the smart objects involved in the encapsulated federation. In order to implement this basic idea, we will introduce membrane objects and path-dependent ports. A membrane object is a smart object that defines a capsule or a compartment for encapsulation. Its role is to encapsulate one of the following three, i.e., a smart object, a service port, or a channel. Smart object encapsulation makes all the ports of the target smart object invisible and inaccessible from any other service-requesting port that is not encapsulated by the same membrane object. Service encapsulation makes the target service-providing port of some smart object invisible and inaccessible from any other service-requesting port that is not encapsulated by the same membrane object. Channel encapsulation makes the target channel invisible from, and non-interceptable by any other smart object. While both smart object encapsulation and service encapsulation can be used to implement object-oriented and/or capability-based access control, channel encapsulation can be used to protect message communication against its interception.

What we need is to define each membrane object as a smart object with some ports. Through the port matching between some of these ports of the membrane smart object and some ports of the target smart object, all the ports of the target smart object or a specified target service-providing port should be securely protected against any nonauthorized access. For channel encapsulation, the membrane smart object should be connected through the port matching with each of the two objects to be connected together with this channel, and protect this channel against the interception of the channel message by any malicious smart object.

In order to implement such membrane objects as smart objects, we will introduce the following two types of guarded ports; $\pm \sigma{:}p$ and $\pm [\sigma]p$, where p and σ are respectively a port and a channel path of non-zero length. They are called guarded ports since the original p ports are guarded respectively by $\sigma{:}$ and $[\sigma]$. They become active under the conditions detailed in the following.

A guarded service-requesting port $-\sigma{:}p$ of a smart object o spans a $\sigma{:}p$ channel from itself to $+\sigma{:}p$ of the smart object $\sigma(o)$, and keeps it active only while the channel path σ is active and $\sigma(o)$ is in the scope of o. Each message passing from $-\sigma{:}p$ to $+\sigma{:}p$ needs to check if the channel path σ is active before sending a message. Otherwise, the message passing is prohibited. A guarded service-providing port $+\sigma{:}p$ in a smart object o is accessible only from a guarded service-requesting port $-\sigma{:}p$ only while $-\sigma{:}p$ is active. Figure 8 shows the rewriting rules for the guardian $\sigma{:}p$. These rules are called path-dependency rules.

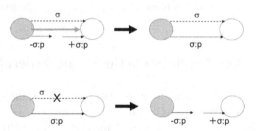

Fig. 8. Graph rewriting rules for guarded ports $\pm \sigma{:}p$.

A guarded port $\pm [\sigma]p$ in a smart object o becomes active and makes itself work as $\pm p$ only while the channel path σ is active and $\sigma(o) = o$. In other words, a guarded service-requesting port $-[\sigma]p$ in a smart object o spans a channel p from itself to $+p$ in another object o' in the scope of o, and keeps it active only while $\sigma(o) = o$. A guarded service-providing port $+[\sigma]p$ becomes accessible from $-p$ in another object o' only while o is in the scope of o' and $\sigma(o) = o$. Figure 9 shows the rewriting rules for the guardian $[\sigma]$. These rules are called loop-path-dependency rules.

A membrane object m_i is a smart object of a special type with two ports, i.e., an entry-service providing port $+me_i$ and a ruled-service requesting port $-mr_i$. Each membrane object m_i provides an entry service me_i for another object o with $-me_i$ to request the

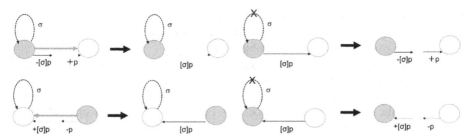

Fig. 9. Graph rewriting rules for guarded ports $\pm$ [σ]p.

registration of o in this entry service me$_i$ so that o can securely access any objects ruled by m$_i$. The membrane object m$_i$, i.e., its entry service me$_i$, may use capability-based and/or object-oriented access control to accept or to reject this registration request, which we will not detail in this paper. This registration is performed through the channel me$_i$ from −me$_i$ of o to +me$_i$ of m$_i$. If the registration request is accepted, then the channel me$_i$ is kept active. Otherwise, this channel is immediately broken.

Each membrane object m$_i$ can rule another object o that has a ruled-service providing port +mr$_i$ by spanning a channel from its −mr$_i$.to the port +mr$_i$ of o. The target object o may accept or reject this ruled-service request based on its capability-based and/or object-oriented access control strategy. If the request is accepted, then the channel mr$_i$ is kept active. Otherwise, it is immediately broken. If the request is accepted, then the target object is put under the protection control by the membrane object, i.e., the membrane object can restrict the access of any service providing port +p of o that are guarded with me$_i$.mr$_i$: as +me$_i$.mr$_i$:p. Once a service providing object o is put under the ruling by a membrane object m$_i$, its service can be accessible only by those objects that are registered to the same membrane object through an me$_i$ channel. An object o is said to have entered the membrane m$_i$ when o has been registered as a service requesting object into m$_i$ or put under the ruling of m$_i$ as a service providing object. Figure 10 shows a membrane object m$_i$ which both a service-providing object o$_1$ and a service-requesting

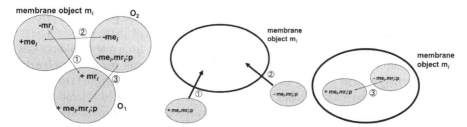

Fig. 10. The leftmost figure shows (1) a ruling request from a membrane object m$_i$ to a service-providing object o$_1$, (2) a registration request from a service-requesting object o$_2$ to the same membrane object m$_i$, and (3) a secure federation between o$_1$ and o$_2$ using the channel me$_i$.mr$_i$:p. The middle figure schematically shows the moment when two objects enter the membrane object, while the rightmost figure schematically shows the secure federation between two objects that are simultaneously encapsulated by the same membrane object.

object o_2 enter to securely federate with each other using a guarded channel $me_i.mr_i{:}p$. The middle and the rightmost figures in Fig. 10 metaphorically represent the membrane object as a compartment which the two objects may enter.

Now we consider how membrane objects and guarded ports can be used to encapsulate objects, services, and/or channels. For the smart object encapsulation, we will encapsulate all the ports of the target smart object. When all the ports of a smart object become invisible, no one can access this object. Figure 11 shows how a smart object o is encapsulated by a membrane object m_i. We assume that the membrane object m_i has a ruled-service requesting port $-mr_i$ and an entry-service providing port $+me_i$ while the target smart object o has a ruled-service providing port $+mr_i$ and an entry-service requesting port $-me_i$, and each of its port $\pm p$ is guarded as $\pm [me_i.mr_i]p$. Without a channel path $me_i.mr_i$ from o to itself through m_i, each port $\pm [me_i.mr_i]p$ in o cannot work as an active port, i.e., it stays invisible and inaccessible, and hence o is encapsulated. In this mechanism, however, o needs to be *a priori* hard-coded under the assumption that it will be encapsulated by a specific membrane object m_i. This problem can be easily solved by introducing additional rewriting rules that can make every port of its σ object guarded with arbitrary guardians. However, this extension may also lower the security level. Therefore, in this paper, we will focus on the hard-coded membrane mechanism.

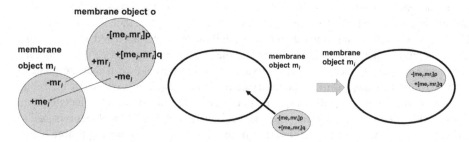

Fig. 11. A membrane object to encapsulate a smart object o.

Once the membrane object m_i makes each port $\pm [me_i.mr_i]p$ of the object o visible as $\pm p$, this port becomes visible also to any other object while the path $me_i.mr_i$ from o to itself through m_i is kept active. Therefore, the encapsulation of all the ports of a target smart object by itself cannot securely protect this smart object from malicious accesses. It can only control the visibility of these ports to other smart objects. In order to solve this problem, we may restrict the arity of $\pm [me_i.mr_i]p$ to one as $\pm [me_i.mr_i]p(1)$ to prohibit any further connection to or from this port after establishing the desired federation using this port. This encapsulation mechanism can be used to break a securely established active federation between o_1 and o_2 from outside, namely by deactivating the encapsulating membrane object m_i. Without m_i, no one can reestablish the same federation between o_1 and o_2.

This mechanism in Fig. 11 is also used in combination with the mechanism in Fig. 10 to strengthen the security protection by the latter. The nested membrane objects use the mechanism of a membrane object m_i in Fig. 11 to control the visibility of all the ports of another membrane object m_j that is used as m_i in Fig. 10 to securely protect a federation from o_1 to o_2. Figure 12 shows this whole mechanism. Without the ruling of m_j by m_i, neither of o_1 or o_2 can enter the membrane m_j to establish their federation.

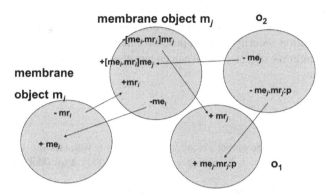

Fig. 12. Nested membrane objects: A membrane object m_i encapsulates another membrane object m_j to securely protect a federation from o_1 to o_2.

The service encapsulation may use the same mechanism as shown in Figs. 10 or 11 only for those service-providing ports of the target smart object o to be encapsulated.

For the channel encapsulation, we may use the mechanism that was already shown in Fig. 10, where the guarded channel $me_i.mr_i$:p is securely encapsulated by a membrane object m_i. For a malicious smart object o' to intercept the message passing through this channel, o' is required to be registered in the membrane object m_i beforehand, and its service-requesting port should be guarded with $me_i.mr_i$: in hard coding.

It should be noted here that the same service p of a smart object o can be simultaneously encapsulated by more than one membrane objects, say, m_i and m_j. For this purpose, o should be hard-coded to have two different guarded ports $+me_i.mr_i$:p and $+me_j.mr_j$:p for the service p. Such hard-coding of guarded ports can protect this object o against the unexpected ruling by a malicious membrane object.

5   Concluding Remarks

Any complex application scenario using the proximity-based federation of a large number of different types of smart objects with wireless communication capabilities can be generally modeled as a binary autocatalytic-reaction network, in which each reaction denotes either federation or defederation of single or composite smart objects. Based on the formal modeling of smart objects proposed by his previous papers, the present author has given a simplified generic way of implementing each catalytic reaction in an arbitrarily given binary autocatalytic-reaction network as a list of additional graph rewriting rules to be coded in each of the involving smart objects. The framework proposed here will open a new vista of novel complex application scenarios of proximity-based federation of smart objects.

The increase of the complexity of application scenarios will necessarily increases the threats of malicious attacks. Because of the highly dynamic and *ad hoc* nature of proximity-based federation of smart objects, complex applications of smart objects may easily become the target of malicious attacks. Secure federation is most important in such applications. This paper has proposed a new framework-level solution to this issue. Our solution is independent from the encryption-based technologies for secure communication between two smart objects, and can be used together with them to increase the federation security. The proposed framework uses a special type of smart objects called membrane objects, which can encapsulate desired smart objects, services, and/or federations.

References

1. Milner, R.: Theories for the global ubiquitous computer. In: Walukiewicz, I. (ed.) FoSSaCS 2004. LNCS, vol. 2987, pp. 5–11. Springer, Heidelberg (2004). https://doi.org/10.1007/978-3-540-24727-2_2
2. Henricksen, K., Indulska, J., Rakotonirainy, A.: Modeling context information in pervasive computing systems. In: Mattern, F., Naghshineh, M. (eds.) Pervasive 2002. LNCS, vol. 2414, pp. 167–180. Springer, Heidelberg (2002). https://doi.org/10.1007/3-540-45866-2_14
3. Tanaka, Y.: Proximity-based federation of smart objects: liberating ubiquitous computing from stereotyped application scenarios. In: Setchi, R., Jordanov, I., Howlett, R.J., Jain, L.C. (eds.) KES 2010. LNCS (LNAI), vol. 6276, pp. 14–30. Springer, Heidelberg (2010). https://doi.org/10.1007/978-3-642-15387-7_6
4. Julia, J., Tanaka, Y.: Proximity-based federation of smart objects: its graph-rewriting framework and correctness. J. Intell. Inf. Syst. **46**(1), 147–178 (2016)
5. Tanaka, Y.: Proximity-based federation of smart objects and their application framework. In: Kyung, C.-M., Yasuura, H., Liu, Y., Lin, Y.-L. (eds.) Smart Sensors and Systems, pp. 411–439. Springer, Cham (2017). https://doi.org/10.1007/978-3-319-33201-7_15
6. Edwards, W.K., Joy, B., Murphy, B.: Core JINI. Prentice Hall Professional Technical Reference, Upper Saddle River (2000)

4-Valued Semantics Under the OWA: A Deductive Database Approach

Dominique Laurent[(✉)]

ETIS Laboratory - ENSEA/UCP/CNRS, Cergy-Pontoise, France
dominique.laurent@u-cergy.fr

Abstract. In this paper, we introduce a novel approach for dealing with databases containing inconsistent information. Considering four-valued logics in the context of OWA (Open World Assumption), a database Δ is a pair (E, R) where E is the extension and R the set of rules. In our formalism, the set E is a set of pairs of the form $\langle \varphi, \mathbf{v} \rangle$ where φ is a fact and $\mathbf{v}$ is either $\mathbf{t}$, or $\mathbf{b}$, or $\mathbf{f}$ (meaning respectively true, inconsistent or false), given that unknown facts are not stored. Moreover the rules extend Datalog^{neg} rules allowing their heads to be a negative atom.

We then define the notion of model of such a database, we show how to compute one particular model called semantics, and we investigate properties of this model. We also show how our approach applies to data integration and we review examples from the literature.

Keywords: Open world assumption · Datalog with negation · Inconsistent database · Database semantics

1 Introduction

In this paper, we present and discuss our preliminary work on a novel approach meant to take into account the needs of many current applications, specifically in the domains of data integration and data warehousing:

1. As for usual Datalog databases [6], in our approach, a database Δ is a pair (E, R) where E (respectively R) is called the *extension* (respectively *set of rules*) of Δ. Whereas in standard approaches E is a set of ground facts meant to be *true*, in our approach E is a set of pairs of the form $\langle \varphi, \mathbf{v} \rangle$ where φ is a ground fact and $\mathbf{v}$ is one of the *three* truth values $\mathbf{t}$ (*true*), $\mathbf{b}$ (*contradictory*) or $\mathbf{f}$ (*false*), meaning that a fact can be either true, or contradictory or false, given that facts not occurring in E are considered *unknown*.
2. Recalling that a *literal* is an atom or the negation of an atom, the rules in R are expressions of the form *head* $\leftarrow$ *body* where *head* is a literal and *body* is a conjunction (denoted as a list) of literals. We notice that such rules generalize standard Datalog^{neg} rules since their head may be a *negated* atom.
3. The database semantics is defined based on the four valued semantics as defined in [12,16]. These semantics reflect the *Open World Assumption*

© Springer Nature Switzerland AG 2020
G. Flouris et al. (Eds.): ISIP 2019, CCIS 1197, pp. 101–116, 2020.
https://doi.org/10.1007/978-3-030-44900-1_7

(OWA), contrary to most database models that assume the Closed World Assumption (CWA) [14]. As in [4], we argue that considering the OWA instead of the CWA is relevant in most applications related with data integration on the Web, for the following intuitive reason: when a fact does not appear in the answer to a query, this does *not* mean that this fact is *false*, but rather that it has not been searched properly. Therefore, in the absence of any other information such a fact will be considered *unknown*, instead of false.

As compared with related work on contradictions in databases, we propose a radically different approach. Indeed, the purpose of previous work dealing with contradictions in databases, is either to define and investigate 'repairs' so as to make the database consistent [1,9], and/or to identify a set of queries whose answer is independent from any contradiction [11]. Instead, we propose an approach in which contradictions can be stored or deduced through rules, and our purpose is not to eliminate or avoid contradictions. Instead, our semantics allows for handling contradictory information as such, thus reflecting real world applications in which true, false, inconsistent and unknown information have to be dealt with, as is the case when data integration is involved. In this context, the main contributions of this work are the following:

1. We define semantics for databases containing contradictions in the context of OWA. To do so, we review the formalism introduced in [3] and further discussed in the literature [2,7,12,16], in particular regarding implication.
2. We study minimality properties of our semantics with respect to set inclusion and to the two orderings induced by four-valued logics (*truth* and *knowledge* orderings). We also relate our semantics to standard Datalog semantics [6].
3. We discuss the relevance of our approach in the context of data integration, a context of interest in more and more current applications. We illustrate this point by reviewing two well-known examples from the literature.

The paper is organized as follows: In Sect. 2 we recall some background on four-valued logics and we introduce basic definitions. Section 3 deals with database semantics and minimality properties. In Sect. 4, we review examples from the literature in the context of our approach. Section 5 concludes the paper.

2   Background: Four-Valued Logics and Database Context

2.1   Basics on Four-Valued Logics

Four-valued logics was introduced by Belnap in [3], who argued that this formalism could be of interest when integrating data from various data sources. To this end, denoting by t, b, n and f the four truth values, the usual connectives $\neg$, $\vee$ and $\wedge$ have been defined as shown in Fig. 1. An important feature of this four-valued logics is that it allows to compare truth values according to two partial orderings, known as *truth ordering* and *knowledge ordering*, respectively denoted by $\preceq_t$ and $\preceq_k$ and defined by:

$$n \preceq_k t \preceq_k b, n \preceq_k f \preceq_k b \quad \text{and} \quad f \preceq_t n \preceq_t t, f \preceq_t b \preceq_t t.$$

As in standard two-valued logics, conjunction (respectively disjunction) corresponds to minimum (respectively maximum) truth value, when considering the truth ordering. It has also been shown in [3,7] that the set $\{t, b, n, f\}$ equipped with these two orderings has a distributive bi-lattice structure. In this bi-lattice, the minimum and maximum with respect to $\preceq_t$ are $\wedge$ and $\vee$, and the minimum and maximum with respect to $\preceq_k$ are respectively denoted by $\otimes$ and $\oplus$.

This four-valued logics has motivated many comments and research, because some basic properties holding in standard logics do not hold in this formalism. For example, Fig. 1 shows that formulas of the form $\Phi \vee \neg\Phi$ are not always true, independently from the truth value of Φ. More importantly, it has been argued in [2,12,16] that defining the implication $\Phi_1 \Rightarrow \Phi_2$ as usual in two-valued logics, that is by $\neg\Phi_1 \vee \Phi_2$, is problematic.

To see this, we consider as in [2,3,12,16], that t and b are the two *designated truth values*, and a formula Φ is said to be *valid* if its truth value is designated. For instance, assuming that the truth value of Φ is n, then $\Phi \Rightarrow \Phi$ is not valid since its truth value is n. Notice in this respect that Fig. 1 shows that $\Phi \to \Phi$, $\Phi \hookrightarrow \Phi$, $\Phi \overset{*}{\to} \Phi$ and $\Phi \overset{*}{\hookrightarrow} \Phi$ are valid for any truth assignment regarding Φ.

As argued in [2,12,16], considering $\Rightarrow$ as an implication does not satisfy the deduction theorem, because the formula Φ defined by $(\Phi_1 \wedge (\Phi_1 \Rightarrow \Phi_2)) \Rightarrow \Phi_2$ is *not* valid for any assignment v such that $v(\Phi_1) = b$ and $v(\Phi_2) = f$ (because in such case, $v(\Phi_1 \Rightarrow \Phi_2) = b$ and thus, $v(\Phi) = f$).

It can however be considered that Modus Ponens holds, in the sense that the statement [*if* $v(\Phi_1)$ *and* $v(\Phi_1 \Rightarrow \Phi_2)$ *are equal to* t *then* $v(\Phi_2) = t$] holds. However, this same statement enhanced by replacing 'true' with 'designated' does not hold because for $v(\Phi_1) = b$ and $v(\Phi_2) = f$ we have $v(\Phi_1 \Rightarrow \Phi_2) = b$. This matter of fact is a severe limitation in our context, which explains why implication defined by '$\Rightarrow$' is discarded as a possible semantics of our rules.

Other implications have been introduced, among which two in [2,12] denoted hereafter by $\to$ and $\overset{*}{\to}$, and two in [16] denoted hereafter by $\hookrightarrow$ and $\overset{*}{\hookrightarrow}$. The truth tables of these implications are shown in Fig. 1.

We recall from [2] (Corollary 9) that $\to$ is defined 'from scratch' in the sense that it cannot be expressed using the other standard connectives $\neg$, $\vee$ and $\wedge$. Moreover, since $\Phi_1 \to \Phi_2$ and $\neg\Phi_2 \to \neg\Phi_1$ are not equivalent, the implication $\Phi_1 \overset{*}{\to} \Phi_2$ is introduced in [2,12] as $(\Phi_1 \to \Phi_2) \wedge (\neg\Phi_2 \to \neg\Phi_1)$. The implication $\Phi_1 \hookrightarrow \Phi_2$ is defined in [16] by $\sim \Phi_1 \vee \Phi_2$, where $\sim$ is a complement operator whose truth table is shown in Fig. 1. Again, $\Phi_1 \hookrightarrow \Phi_2$ and $\neg\Phi_2 \hookrightarrow \neg\Phi_1$ are not equivalent, and so, $\Phi_1 \overset{*}{\hookrightarrow} \Phi_2$ is defined as $(\Phi_1 \hookrightarrow \Phi_2) \wedge (\neg\Phi_2 \hookrightarrow \neg\Phi_1)$.

Since the formula Φ defined by $(\Phi_1 \wedge (\Phi_1 \rightsquigarrow \Phi_2)) \rightsquigarrow \Phi_2$ is valid for any truth value assignment when replacing $\rightsquigarrow$ with one of the implications $\to$, $\hookrightarrow$, $\overset{*}{\to}$ or $\overset{*}{\hookrightarrow}$, Modus Ponens does apply in its standard *and* enhanced forms for any of these implications. It is also interesting to see that when merging the truth values t and b (respectively f and n) into a single truth value, say True (respectively False), the corresponding truth tables of $\to$ and $\hookrightarrow$ are that of the standard implication, while this is not the case for $\Rightarrow$, $\overset{*}{\to}$ and $\overset{*}{\hookrightarrow}$. This explains why we

φ	$\neg\varphi$
t	f
b	b
n	n
f	t

φ	$\sim\varphi$
t	f
b	n
n	b
f	t

$\vee$	t	b	n	f
t	t	t	t	t
b	t	b	t	b
n	t	t	n	n
f	t	b	n	f

$\wedge$	t	b	n	f
t	t	b	n	f
b	b	b	f	f
n	n	f	n	f
f	f	f	f	f

$\Rightarrow$	t	b	n	f
t	t	b	n	f
b	t	b	t	b
n	t	t	n	n
f	t	t	t	t

$\rightarrow$	t	b	n	f
t	t	b	n	f
b	t	b	n	f
n	t	t	t	t
f	t	t	t	t

$\hookrightarrow$	t	b	n	f
t	t	b	n	f
b	t	t	n	n
n	t	b	t	b
f	t	t	t	t

$\overset{*}{\rightarrow}$	t	b	n	f
t	t	f	n	f
b	t	b	n	f
n	t	n	t	n
f	t	t	t	t

$\overset{*}{\hookrightarrow}$	t	b	n	f
t	t	f	f	f
b	t	t	f	f
n	t	f	t	f
f	t	t	t	t

Fig. 1. Truth tables of basic connectors and implications

discard these three implications. However, the choice between $\rightarrow$ and $\hookrightarrow$ is not easy for the following reasons:

- In [2,12], it is argued that, similarly to two-valued implication, $\rightarrow$ satisfies the property that $v(\Phi_1 \rightarrow \Phi_2) = v(\Phi_2)$ whenever $v(\Phi_1)$ is designated. However, $\rightarrow$ does not satisfy the properties of $\hookrightarrow$ given below.
- Although $\hookrightarrow$ does not satisfy the above property, it is argued in [16] that, similarly to two-valued implication, $\hookrightarrow$ satisfies the property that $v(\Phi_1) \preceq_t v(\Phi_2)$ if and only if $v(\Phi_1 \hookrightarrow \Phi_2) = \mathsf{t}$.

We draw attention on that none of these two implications does satisfy all intuitively appealing properties that standard two-valued implication satisfies, among which contraposition is an important example. Since in our approach, implications are seen as rules, Modus Ponens is the basic 'logical tool' to be used, whereas contraposition is not used as such.

Looking at the truth tables of the two implications $\rightarrow$ and $\hookrightarrow$ shown in Fig. 1, when the left hand side is valid in S, it is necessary that the right hand side be also valid in order to make the implication valid. More precisely, if Φ_1 is valid, the implications $\Phi_1 \rightarrow \Phi_2$ and $\Phi_1 \hookrightarrow \Phi_2$ are valid in S for any truth assignment v such that:

- $v(\Phi_1) = \mathsf{t}$ and $v(\Phi_2) = \mathsf{t}$ or $v(\Phi_2) = \mathsf{b}$,
- $v(\Phi_1) = \mathsf{b}$ and $v(\Phi_2) = \mathsf{t}$ or $v(\Phi_2) = \mathsf{b}$.

As a consequence, if it happens that Φ_1 is valid while Φ_2 is not, the implication can be made valid by changing the truth value of Φ_2 in two ways: making it either true or inconsistent. As will be seen later, we choose to set $v_S(\Phi_2)$ as equal to $v_S(\Phi_1)$. This choice is motivated by the fact that it is the only one satisfying $v(\Phi_1) \preceq_k v(\Phi_2)$ and $v(\Phi_1) \preceq_t v(\Phi_2)$.

2.2 Four-Valued Logics in the Database Context

As usual when dealing with deductive databases, the considered alphabet is made of constants, variables and predicate symbols, and ground atoms are called facts.

We thus assume a fixed set of facts, called *universe* and denoted by $\mathcal{U}$. The set $\mathcal{U}$ is the set of all possible facts that can be built up using the constants and predicates that occur in the database being modeled. It is therefore important to notice that, as databases are assumed to be finite, so is $\mathcal{U}$.

While in the two-valued setting under the CWA, the database extension and the database semantics are sets of facts, meant to be true and the facts not occurring in the database semantics are set to be false, in our context of four-valued logics under the OWA, the database extension and the database semantics may contain facts that are either true, contradictory or false, assuming that non stored facts are unknown.

To account for this situation, we consider sets of pairs $\langle \varphi, \mathbf{v} \rangle$ where φ is a fact in $\mathcal{U}$ and where $\mathbf{v}$ is one of the truth values $\mathbf{t}, \mathbf{b}, \mathbf{n}$ or $\mathbf{f}$, and we assume that facts whose truth value is $\mathbf{n}$ are not stored. Such a set S is said to be *consistent* if for all distinct pairs $\langle \varphi_1, \mathbf{v}_1 \rangle$ and $\langle \varphi_2, \mathbf{v}_2 \rangle$ in S, $\varphi_1 \neq \varphi_2$. Consequently a consistent set S is seen as a valuation v_S defined for every φ in $\mathcal{U}$ by:

$$v_S(\varphi) = \mathbf{v}, \text{ if } S \text{ contains a pair } \langle \varphi, \mathbf{v} \rangle; v_S(\varphi) = \mathbf{n}, \text{ otherwise.}$$

Consistent sets of pairs are called *v-sets*, standing for *valuated* sets. It should be noticed that, in [16], a formal way of expressing the pairs in a v-set is proposed by defining unary operators. We however do not use this formalism because its proper definition requires further notation which could be thought unnecessarily sophisticated in our context.

Given a v-set S and a formula Φ, based on the truth tables given in Fig. 1, Φ is said to be *valid in* S if $v_S(\Phi)$ is designated. For example, $a \to b$ is valid in $S_1 = \{\langle a, \mathbf{t} \rangle, \langle b, \mathbf{b} \rangle\}$ because $v_{S_1}(a \to b) = \mathbf{b}$, but $a \to b$ is not valid in $S_2 = \{\langle a, \mathbf{t} \rangle\}$ because $v_{S_2}(a \to b) = \mathbf{n}$.

The two orderings $\preceq_k$ and $\preceq_t$ are extended to v-sets over the same universe $\mathcal{U}$ as follows.

Definition 1. *For all v-sets S_1 and S_2 over $\mathcal{U}$, $S_1 \preceq_k S_2$, respectively $S_1 \preceq_t S_2$, holds if for every φ in $\mathcal{U}$, $v_{S_1}(\varphi) \preceq_k v_{S_2}(\varphi)$, respectively $v_{S_1}(\varphi) \preceq_t v_{S_2}(\varphi)$, holds.*

For example with $\mathcal{U} = \{a, b, c\}$, $S_1 = \{\langle a, \mathbf{t} \rangle\}$ and $S_2 = \{\langle a, \mathbf{b} \rangle, \langle b, \mathbf{f} \rangle\}$, we have $v_{S_1}(b) = v_{S_1}(c) = v_{S_2}(c) = \mathbf{n}$ and so:

- $v_{S_1}(a) \preceq_k v_{S_2}(a)$, $v_{S_1}(b) \preceq_k v_{S_2}(b)$ and $v_{S_1}(c) \preceq_k v_{S_2}(c)$, implying that $S_1 \preceq_k S_2$ holds.
- $v_{S_2}(a) \preceq_t v_{S_1}(a)$, $v_{S_2}(b) \preceq_t v_{S_1}(b)$ and $v_{S_2}(c) \preceq_t v_{S_1}(c)$, implying that $S_2 \preceq_t S_1$ holds.
- $\emptyset \preceq_k S_2$, because for every φ, $v_\emptyset(\varphi) = \mathbf{n}$, the least value with respect to $\preceq_k$.
- $\emptyset$ and S_2 are not comparable with respect to $\preceq_t$, because $v_\emptyset(a) = \mathbf{n}$ and $v_{S_2}(a) = \mathbf{b}$ are not comparable with respect to $\preceq_t$.

The extension of $\preceq_k$ generalizes set inclusion in the sense that if $S_1 \subseteq S_2$, then we have $S_1 \preceq_k S_2$. Notice that, as the last item above shows, the truth ordering $\preceq_t$ does not satisfy this property, because $\emptyset \subseteq S_2$ holds while $\emptyset \preceq_t S_2$ does not.

3   Database and Database Semantics

As in standard approaches to Datalog databases [5,6], a database consists of an *extension* and a *set of rules*, formally defined as follows.

Definition 2. *A database Δ is a pair $\Delta = (E, R)$ where E and R are respectively called the* extension *and the* rule set *of Δ. If $\Delta = (E, R)$, then:*

- *E is a v-set.*
- *R is a set of rules of the form $\rho : h \leftarrow b_1, \ldots, b_n$ where*
  1. *for $i = 1, \ldots, n$, b_i is a literal (positive or negative) and the set of all b_i's $(i = 1, \ldots, n)$ is called the* body *of ρ, denoted by $body(\rho)$,*
  2. *h is a positive or negative literal, called the* head *of ρ, denoted by $head(\rho)$,*
  3. *all variables occurring in h also occur in $body(\rho)$, i.e., rules are safe.*

It is important to notice that the rules in our approach generalize Datalog^{neg} rules [5] because negative atoms are allowed not only in the body but also in the head of the rules. This implies that rules may generate contradictory facts.

As usual, rules are seen as implications, either $\rightarrow$ or $\hookrightarrow$ that must be valid in the database semantics. Notice in this respect that Fig. 1 shows that for all formulas ϕ_1 and ϕ_2, $\phi_1 \rightarrow \phi_2$ is valid if and only if so is $\phi_1 \hookrightarrow \phi_2$. This explains why our approach can be said 'compatible' with either implication.

Similarly to the standard Datalog approach, a model of a database $\Delta = (E, R)$ could be defined as a v-set M containing E and in which all rules in R are valid. However, such a definition would raise important problems:

1. *A database might have no model.* To see this, consider $\Delta = (E, R)$ where $R = \{b \leftarrow a\}$ and where $E = \{\langle a, \mathtt{t}\rangle, \langle b, \mathtt{f}\rangle\}$. Whatever the chosen implication (either $\rightarrow$ or $\hookrightarrow$), in any model M, $v_M(a \rightarrow b) = v_M(a \hookrightarrow b) = \mathtt{f}$ because M must contain the two pairs of E. Notice that this cannot happen in standard Datalog since the storage of false facts is not allowed.
2. *A database might have more than one minimal model, with respect to set inclusion.* This case is illustrated above where $S_1' = \{\langle a, \mathtt{t}\rangle, \langle b, \mathtt{t}\rangle\}$ are $S_2' = \{\langle a, \mathtt{t}\rangle, \langle b, \mathtt{b}\rangle\}$ two minimal v-sets containing $\{\langle a, \mathtt{t}\rangle\}$ in which $b \leftarrow a$ is valid. This situation does not happen in standard Datalog because the minimal model is known to be unique.

Whereas the second issue raised above will be further investigated later, the first issue is solved in our approach by giving the priority to the database extension over the rules. To do so, we prevent from applying a rule in R when it leads to some conflict with a pair in E. In order to implement this policy, given a database $\Delta = (E, R)$ over universe $\mathcal{U}$, we denote by $inst(E, R)$ the set of all instantiations ρ of rules in R such that $head(\rho)$ does not occur in E. Moreover, given a rule $\rho : h \leftarrow b$ we denote by $\rho^{\rightarrow}$, respectively $\rho^{\hookrightarrow}$, the formula $b \rightarrow h$, respectively $b \hookrightarrow h$. The definition of a model of Δ then follows.

Definition 3. *Let $\Delta = (E, R)$ be a database over universe $\mathcal{U}$. A v-set M is a* model *of Δ if the following holds:*

1. $E \subseteq M$, *i.e.*, M *must contain the database extension, and*
2. *every* ρ *of* $inst(E, R)$ *is valid in* M, *i.e.*, $v_M(\rho^{\rightarrow})$ *and* $v_M(\rho^{\leftrightarrow})$ *are designated.*

Referring to the previous two items, Definition 3 applies as follows:

- For $\Delta = (E, R)$ with $E = \{\langle a, \mathbf{t} \rangle, \langle b, \mathbf{f} \rangle\}$ and $R = \{b \leftarrow a\}$, E is a model of Δ since we have $inst(E, R) = \emptyset$. Notice that here, E is the only minimal model with respect to set inclusion.
- For $\Delta = (E, R)$ with $E = \{\langle a, \mathbf{t} \rangle\}$ and $R = \{b \leftarrow a\}$, $S_1' = \{\langle a, \mathbf{t} \rangle, \langle b, \mathbf{t} \rangle\}$ are $S_2' = \{\langle a, \mathbf{t} \rangle, \langle b, \mathbf{b} \rangle\}$ two minimal models of Δ, because $inst(E, R) = R$ and the two v-sets satisfy Definition 3.

Given a database Δ, a modified version of the membership immediate consequence operator [5,7] is defined below. It will then be seen that this allows for computing a particular model of Δ, which we call the *semantics* of Δ.

Definition 4. *Let* $\Delta = (E, R)$ *be a database. The semantic consequence operator associated with* Δ, *denoted by* Σ_Δ, *is defined for every v-set* S *by the following steps:*

(1) Define first $\Gamma_\Delta^E(S)$ *as follows:*

$$
\begin{aligned}
\Gamma_\Delta^E(S) = E \;\cup\; & \{\langle h, \mathbf{t} \rangle \mid (\exists \rho \in inst(E, R))(h = head(\rho) \wedge v_S(body(\rho)) = \mathbf{t})\} \\
\cup\; & \{\langle h, \mathbf{b} \rangle \mid (\exists \rho \in inst(E, R))(h = head(\rho) \wedge v_S(body(\rho)) = \mathbf{b})\} \\
\cup\; & \{\langle h, \mathbf{f} \rangle \mid (\exists \rho \in inst(E, R))(\neg h = head(\rho) \wedge v_S(body(\rho)) = \mathbf{t})\} \\
\cup\; & \{\langle h, \mathbf{b} \rangle \mid (\exists \rho \in inst(E, R))(\neg h = head(\rho) \wedge v_S(body(\rho)) = \mathbf{b})\}
\end{aligned}
$$

(2) Then, $\Sigma_\Delta(S)$ *is defined by:*

$$
\Sigma_\Delta(S) = \{\langle \varphi, v_\oplus(\varphi) \rangle \mid \varphi \text{ occurs in } \Gamma_\Delta^E(S)\}
$$

where $v_\oplus(\varphi) = \max_k\{v \mid \langle \varphi, v \rangle \in \Gamma_\Delta^E(S)\}$.

The following lemma, whose proof can found in Appendix A, shows basic properties of the operator Σ_Δ.

Lemma 1. *For every* $\Delta = (E, R)$ *and all v-sets* S, S_1 *and* S_2:

1. $\Sigma_\Delta(S)$ *is a v-set such that* $E \preceq_k \Sigma_\Delta(S)$.
2. *If* $S_1 \preceq_k S_2$, *then* $\Sigma_\Delta(S_1) \preceq_k \Sigma_\Delta(S_2)$.

As a consequence of Lemma 1, with respect to $\preceq_k$, the sequence defined by

- $\Sigma^0 = E$
- for every $n \geq 1$, $\Sigma^n = \Sigma_\Delta(\Sigma^{n-1})$

is monotonic, that is for every $i \geq 0$, $\Sigma^i \preceq_k \Sigma^{i+1}$, and since $\mathcal{U}$ is finite, this sequence has a unique limit. We denote this limit as Σ_Δ^* and we call it the *semantics* of Δ. Moreover, the valuation $v_{\Sigma_\Delta^*}$ is more simply denoted by v_Δ and the valid facts in Σ_Δ^* are said to be *valid in* Δ.

As a consequence of the above properties, it should be clear that Σ_Δ^* is a v-set such that $E \subseteq \Sigma_\Delta^*$ and $E \preceq_k \Sigma_\Delta^*$. This intuitively means that the semantics *extends* the content and the knowledge provided by the database extension E. The following example illustrates cases of computation of Σ_Δ^*.

Example 1. Consider the database $\Delta = (E, R)$ over $\mathcal{U} = \{a, b, c, d, e\}$ where R is the set of the following three rules

$$\rho_1 : c \leftarrow a, \neg b; \quad \rho_2 : e \leftarrow d; \quad \rho_3 : c \leftarrow e$$

and where $E = \{\langle a, \mathtt{t}\rangle, \langle b, \mathtt{f}\rangle, \langle e, \mathtt{b}\rangle\}$. In this case, $inst(E, R) = \{\rho_1, \rho_3\}$ and the computation of $\Sigma_{\Delta'}^*$ is as follows:

1. $\Sigma^1 = E$.
2. $\Sigma^1 = \Sigma_{\Delta'}(\Sigma^0)$. We have $\Gamma_{\Delta'}^\in(\Sigma^0) = E \cup \{\langle c, \mathtt{t}\rangle, \langle c, \mathtt{b}\rangle\}$, which is clearly not an acceptable result since it is not a proper v-set. When computing Σ^1, the first pair is removed, and we obtain that $\Sigma^1 = E \cup \{\langle c, \mathtt{b}\rangle\}$.
3. Since $\Sigma^2 = \Sigma^1$ (no rule applies on Σ^1 to produce new pairs), the computation stops returning that $\Sigma_\Delta^* = \Sigma^1 = \{\langle a, \mathtt{t}\rangle, \langle b, \mathtt{f}\rangle, \langle c, \mathtt{b}\rangle, \langle e, \mathtt{b}\rangle\}$.

We draw attention on that E and Σ_Δ are not comparable with respect to $\preceq_t$ because $v_E(c) = \mathtt{n}$ and $v_\Delta(c) = \mathtt{b}$. Hence, $E \preceq_t \Sigma_\Delta$ does not hold in general, contrary to $E \preceq_k \Sigma_\Delta$. Moreover, Σ_Δ^* is a model of Δ because $E \subseteq \Sigma_\Delta^*$ and:

- $\rho_1^\rightarrow$ and $\rho_1^\leftrightarrow$ are valid in Σ_Δ^* because $v_\Delta(\rho_1^\rightarrow) = v_\Delta(\rho_1^\leftrightarrow) = \mathtt{b}$, due to the fact that $v_\Delta(a \wedge \neg b) = \mathtt{t}$ and $v_\Delta(c) = \mathtt{b}$.
- $\rho_2^\rightarrow$ and $\rho_2^\leftrightarrow$ are also valid in Σ_Δ^*. Indeed, as $v_\Delta(d) = \mathtt{n}$ and $v_\Delta(e) = \mathtt{b}$, we have $v_\Delta(\rho_2^\rightarrow) = \mathtt{t}$ and $v_\Delta(\rho_2^\leftrightarrow) = \mathtt{b}$.
- $\rho_3^\rightarrow$ and $\rho_3^\leftrightarrow$ are valid in Σ_Δ^* as well. Indeed, as $v_\Delta(e) = v_\Delta(c) = \mathtt{b}$, we have $v_\Delta(\rho_3^\rightarrow) = \mathtt{b}$ and $v_\Delta(\rho_3^\leftrightarrow) = \mathtt{t}$.   □

The following proposition, whose proof can be found in Appendix B, shows that, as seen in Example 1, Σ_Δ^* is a minimal model of Δ with respect to set inclusion.

Proposition 1. *Given a database $\Delta = (E, R)$, Σ_Δ^* is a minimal (with respect to set inclusion) model of Δ.*

However, the following example shows that Σ_Δ^* is *not* the only minimal model with respect to set inclusion and that Σ_Δ^* is neither minimal nor maximal with respect to any of the two orderings $\preceq_k$ and $\preceq_t$.

Example 2. Considering, as in Example 1, $\Delta = (E, R)$ where $E = \{\langle a, \mathtt{t}\rangle, \langle b, \mathtt{f}\rangle, \langle e, \mathtt{b}\rangle\}$ and $R = \{\rho_1, \rho_2, \rho_3\}$, we recall that $\Sigma_\Delta^* = \{\langle a, \mathtt{t}\rangle, \langle b, \mathtt{f}\rangle, \langle c, \mathtt{b}\rangle, \langle e, \mathtt{b}\rangle\}$. Let S and S' be the following v-sets:

$$S = \{\langle a, \mathtt{t}\rangle, \langle b, \mathtt{f}\rangle, \langle c, \mathtt{t}\rangle, \langle e, \mathtt{b}\rangle\} \quad \text{and} \quad S' = \{\langle a, \mathtt{t}\rangle, \langle b, \mathtt{f}\rangle, \langle c, \mathtt{b}\rangle, \langle d, \mathtt{f}\rangle, \langle e, \mathtt{b}\rangle\}.$$

We first show that S and S' are two models of Δ. Indeed:

- $E \subseteq S$ and $E \subseteq S'$.
- $\rho_1^\rightarrow$ and $\rho_1^\leftrightarrow$ are valid in S and S' because $v_S(\rho_1^\rightarrow) = v_S(\rho_1^\leftrightarrow) = v_{S'}(\rho_1^\rightarrow) = v_{S'}(\rho_1^\leftrightarrow) = \mathtt{t}$, since $v_S(a \wedge \neg b) = v_{S'}(a \wedge \neg b) = v_S(c) = v_{S'}(c) = \mathtt{t}$.
- $\rho_2^\rightarrow$ and $\rho_2^\leftrightarrow$ are not tested since ρ_2 is not in $inst(E, R)$.

- $\rho_3^{\rightarrow}$ and $\rho_3^{\leftrightarrow}$ are valid in S and in S'. Indeed, as $v_S(e) = \mathtt{b}$ and $v_S(c) = \mathtt{t}$, we have $v_S(\rho_3^{\rightarrow}) = v_S(\rho_3^{\leftrightarrow}) = \mathtt{t}$, and since $v_{S'}(e) = \mathtt{b}$ and $v_{S'}(c) = \mathtt{b}$, we have $v_{S'}(\rho_3^{\rightarrow}) = \mathtt{b}$ and $v_{S'}(\rho_3^{\leftrightarrow}) = \mathtt{t}$.

It can be seen that Σ_Δ is *not* the unique minimal model of Δ with respect to set inclusion, because S is also such a minimal model, due to the fact that no proper subset of S does satisfy Definition 3. Notice also that S' is not minimal with respect to set inclusion, because $\Sigma_\Delta^* \subset S'$ holds.

Regarding minimality or maximality of Σ_Δ^* with respect to $\preceq_t$ or $\preceq_k$, we emphasize the following:

- We have $S \prec_k \Sigma_\Delta^*$ and $\Sigma_\Delta^* \prec_t S$, because for every φ in $\mathcal{U}$ different from c, $v_S(\varphi) = v_\Delta(\varphi)$, $v_S(c) \prec_k v_\Delta(c)$, and $v_\Delta(c) \prec_t v_S(c)$. Therefore, Σ_Δ^* is not minimal with respect to $\prec_k$ and not maximal with respect to $\prec_t$.
- We have $S' \prec_t \Sigma_\Delta^*$ and $\Sigma_\Delta^* \prec_k S'$, because for every φ in $\mathcal{U}$ different from d, $v_{S'}(\varphi) = v_\Delta(\varphi)$ and $v_{S'}(d) = \mathtt{f}$, $v_\Delta(d) = \mathtt{n}$, implying that $v_{S'}(d) \prec_t v_\Delta(d)$ and that $v_\Delta(d) \prec_k v_{S'}(d)$. Therefore, Σ_Δ^* is not minimal with respect to $\prec_t$ and not maximal with respect to $\prec_k$. □

If $\Delta = (E, R)$ is a Datalog database, *i.e.*, all facts are associated with $\mathtt{t}$ in E and all literals in the rules in R are positive, then it is easy to see that $\Sigma_\Delta^* = \{\langle\varphi, \mathtt{t}\rangle \mid \varphi \in M_2\}$, where M_2 is the unique minimal model of Δ, as computed in Datalog. Thus, the true facts in Δ are the same as those, when considering Δ as a Datalog database. Notice however that for any φ *not* in M_2, φ is *false* in the Datalog approach, and φ is *unknown* in our approach.

Considering now the more generic case where the rules are Datalog^{neg} rules and where false facts are allowed in E, the following proposition, whose proof is shown in Appendix C, shows that in this case, all minimal models of Δ have the *same* valid facts and the same false facts. To this end, given a v-set S, we denote by $\mathsf{V}(S)$, respectively $\mathsf{F}(S)$, the set of all facts φ that are valid in S (*i.e.*, such that $v_S(\varphi) = \mathtt{t}$ or $v_S(\varphi) = \mathtt{b}$), respectively false in S (*i.e.*, such that $v_S(\varphi) = \mathtt{f}$).

Proposition 2. *Let $\Delta = (E, R)$ be such that for every rule ρ in R, $head(\rho)$ is a positive literal. For all minimal models M_1 and M_2 of Δ, the following holds: (i) $\mathsf{V}(M_1) = \mathsf{V}(M_2)$ and (ii) $\mathsf{F}(M_1) = \mathsf{F}(M_2)$.*

4 Application to Data Integration

4.1 The Generic Scenario

Data integration is a generic context where our approach can be found useful. Indeed, assuming that p data sources, *i.e.*, p databases, are to be integrated in one database, it may happen that two distinct data sources contain contradictory pieces of information. Formally, the integrated database extension is defined by a valuation v defined as follows for every fact φ:

- $v(\varphi) = \mathtt{t}$ if φ is true in all sources providing information about φ.
- $v(\varphi) = \mathtt{b}$ if φ is true in some sources and false in some other sources.

- $v(\varphi) = \mathbf{f}$ if φ is false in all sources providing information about φ.
- $v(\varphi) = \mathbf{n}$ if no data sources provide information about φ.

In other words, assuming that for $i = 1, \ldots, p$, the extension of the ith data source is defined by a valuation v_i, the valuation v defining the integrated database is defined for every fact φ by: $v(\varphi) = \max_k\{v_i(\varphi) \mid i = 1, \ldots, p\}$. Next, we review two examples from the literature in the framework of this scenario.

4.2   The Case of Nixon Diamond

This example deals with the following: on the one hand, quakers are known to be doves and republican are known to be hawks, and on the other hand, being a dove is not compatible with being a hawk. Knowing that Nixon (President of the USA, in the seventies) was a quaker and a republican, the question is: *Was Nixon a hawk or a dove?* The corresponding 'program' is as follows:

$$dove(x) \leftarrow quaker(x) \; ; \; hawk(x) \leftarrow republican(x)$$
$$\neg dove(x) \leftarrow hawk(x) \; ; \; \neg hawk(x) \leftarrow dove(x)$$
$$quaker(Nixon) \; ; \; republican(Nixon)$$

Notice that this can not be seen as a Datalog^{neg} program, because of the third and fourth rules. In the literature this program has been the subject of many comments, and one of the most common approaches is known as *stable semantics* [8]. According to this approach, two minimal models are found: one asserting that Nixon is a quaker and thus a dove, and the other one asserting that Nixon is a republican and thus a hawk.

In our approach, this can be thought of as stemming from two sources $S_1 = \{\langle quaker(Nixon), \mathbf{t}\rangle\}$ and $S_2 = \{\langle republican(Nixon), \mathbf{t}\rangle\}$. According to the above generic scenario, the integrated database $\Delta = (E, R)$ is such that $E = S_1 \cup S_2$ and R contains the four rules displayed above. It is then easy to see that Σ_Δ contains $\langle dove(Nixon), \mathbf{b}\rangle$ and $\langle hawk(Nixon), \mathbf{b}\rangle$, meaning that $dove(Nixon)$ and $hawk(Nixon)$ are contradictory.

A slightly different scenario of integration is to consider that the sources are aware of the rules and thus that they are defined by

- $S_1' = \{\langle quaker(Nixon), \mathbf{t}\rangle, \langle dove(Nixon), \mathbf{t}\rangle, \langle hawk(Nixon), \mathbf{f}\rangle\}$ and
- $S_2' = \{\langle republican(Nixon), \mathbf{t}\rangle, \langle hawk(Nixon), \mathbf{t}\rangle, \langle dove(Nixon), \mathbf{f}\rangle\}$.

The integrated database $\Delta' = (E', R)$ is such that $E' = \{\langle quaker(Nixon), \mathbf{t}\rangle, \langle republican(Nixon), \mathbf{t}\rangle, \langle hawk(Nixon), \mathbf{b}\rangle, \langle dove(Nixon), \mathbf{b}\rangle\}$, which leads as above to the fact that $dove(Nixon)$ and $hawk(Nixon)$ are contradictory.

As an extension to this example, some authors additionally consider that hawks and doves are politically motivated, and study the question: *Is Nixon politically motivated?* Formally two rules are added to the above program, namely $pm(x) \leftarrow dove(x)$ and $pm(x) \leftarrow hawk(x)$, and considering stable semantics, as $pm(Nixon)$ holds in each minimal model, it is concluded that this fact holds.

In our approach, according to our generic scenario, and assuming that the sources are not aware of the rules, $\langle pm(Nixon), \mathbf{b}\rangle$ belongs to the semantics of the integrated database, because $\langle dove(Nixon), \mathbf{b} \rangle$ and $\langle hawk(Nixon), \mathbf{b} \rangle$ are also in the semantics of this database.

However, if we now assume that the sources are aware of the rules, then they are defined by $S_1'' = S_1' \cup \{\langle pm(Nixon), \mathbf{t} \rangle\}$ and $S_2'' = S_2' \cup \{\langle pm(Nixon), \mathbf{t} \rangle\}$. Then, contrary to the previous case, $\langle pm(Nixon), \mathbf{t} \rangle$ belongs to the semantics of the integrated database $\Delta'' = (E'', R)$, because this pair is in E''.

This example shows that, when integrating data in a deductive context, deduction and integration do *not* commute in general. A generic scenario in this respect would assume that each data source has its own facts and rules, whereas the integration site has its own rules. Then, in this context, data integration can be processed according to the following distinct policies:

1. Data and rules are collected from each source and deductions are computed in the integration site using the integrated data and all rules obtained from the data sources.
2. Only data are collected from data sources, assuming that each data source has applied its own rules beforehand. Then, further deductions are computed to the integrated data, using the rules available in the integration site.

Although it is likely that these policies give different results, we notice that the second one is computationally easier for the integration site. It is thus relevant to identify cases where commutativity (or a weak but acceptable form of commutativity) holds, a basic issue left to future work.

4.3   The Trial Example of [13]

This example is about integrating data coming from two sources meant to represent respectively a prosecutor and a lawyer. Given a person λ put on trial, both sources provide information to the judge, the prosecutor trying to convince the judge that λ is suspect and the lawyer trying to convince the judge that λ is innocent. The judge then integrates this information and decides on charging or not λ. A simplified version of this process is expressed in [13] as follows.

$$suspect(x) \leftarrow motive(x) \vee witness(x)$$
$$innocent(x) \leftarrow alibi(x, y) \wedge \neg friends(x, y)$$
$$friends(x, y) \leftarrow friends(y, x) \vee (friends(x, z) \wedge friends(z, y))$$
$$charge(x) \leftarrow suspect(x) \oplus \neg innocent(x)$$

Here, the syntax for the rules is that of Fitting programs [7] where the connectors $\vee$, $\wedge$, $\oplus$ (maximum with respect to $\preceq_k$) and $\otimes$ (minimum with respect to $\preceq_k$) appear in the bodies of the rules. This syntax is compatible with our approach because (*i*) it is well known that the use of $\vee$ allows to group all rules with the same head into one rule, and (*ii*) allowing the connectors $\oplus$ and $\otimes$ in our approach is a sound extension, that is left to future work.

To clarify the intuitive meaning of the last rule, we explain how the judge decides on charging a person λ:

– If either $suspect(\lambda)$ or $\neg innocent(\lambda)$ is true while the other is neither false nor contradictory, then λ is charged.

- If $suspect(\lambda)$ and $\neg innocent(\lambda)$ are contradictory (one is true and the other is false, or one is contradictory), then $charge(\lambda)$ is also contradictory, meaning that the trial needs to be refined, because one of the sources is lying...
- If these two formulas are unknown, then so is also the fact that λ is charged. In this case more information is needed to reach a conclusion.

As in [13], but in our formalism, assume that the prosecutor and the lawyer respectively assert that $\langle witness(John), \mathtt{f} \rangle$ and $\langle friends(John, Ted), \mathtt{t} \rangle$ hold. According to our integration scenario, in the integrated database $\Delta = (E, R)$, E contains these two pairs, thus implying that $charge(John)$ is unknown in Δ.

In [13], the authors show how additional knowledge, called hypotheses, can be checked against such database. For example, it is shown that assuming that $innocent(John)$ is true and that $charge(John)$ is false constitute such a compatible set, and so, that not charging John is consistent.

Although, in our approach, the goal is different, since we look for deriving information from the given program, independently from any hypothesis, the work of [13] can be stated by: *given $\Delta = (E, R)$ and hypotheses H (as a set of pairs), does Δ have a model containing H?* For instance in our case, with $H = \{\langle innocent(John), \mathtt{t} \rangle, \langle charge(John), \mathtt{f} \rangle\}$, it can be seen that the answer is yes, thus leading to the same conclusion as in [13].

On the other hand, as our approach assumes OWA, no false facts can be derived, unless explicitly stated in the program. We point out in this respect that assuming CWA is not appropriate in the present context. Indeed, assume that $alibi(John, Ted)$ is stored as true and that nothing is known about John's friends. In this case, CWA allows to state that $\neg friends(John, Ted)$ holds, thus that $innocent(John)$ holds as well. However, in order to convince the judge that $innocent(John)$ holds, the lawyer is meant to provide *evidence* of the fact that $friends(John, Ted)$ is false, which is not modeled in [13].

5   Conclusion

In this paper we have introduced a novel approach to deductive databases dealing with contradictory information. This work is motivated by the facts that (*i*) many contradictions occur in the real world, and (*ii*) data integration is a field where such contradictions are common. We thus consider a deductive database approach based on the four-valued logics introduced in [3]. Our database semantics has been shown to be 'compatible' with two popular implications [2,16], and some basic properties regarding minimality have been investigated.

As this work is preliminary, many issues are still to be investigated: (*i*) consider extended rules as defined in [7]; (*ii*) investigate the use of the two-valued operators introduced in [16] to allow rules expressing sentences such as *if φ is unknown then φ' is false*; (*iii*) when rules are Datalog^{neg} rules, compare our semantics with those of Datalog^{neg} [5]; (*iv*) integrate standard constraints such as functional dependencies in our approach; (*v*) design and study a relational algebra dealing with tuples in the four-valued logics framework; and (*vi*) define a measurement of inconsistency, as in previous work such as [10] or [15].

A   Proof of Lemma 1

Lemma 1. *For every $\Delta = (E, R)$ and all v-sets S, S_1 and S_2:*

1. *$\Sigma_\Delta(S)$ is a v-set such that $E \preceq_k \Sigma_\Delta(S)$.*
2. *If $S_1 \preceq_k S_2$, then $\Sigma_\Delta(S_1) \preceq_k \Sigma_\Delta(S_2)$.*

Proof. 1. $\Sigma_\Delta(S)$ is a v-set because $v_\oplus(\varphi)$ is unique, for a fixed φ. Moreover, every $\langle \varphi, v \rangle$ in E also belongs to $\Gamma_\Delta^E(S)$, and as no rule can generate another pair involving φ, $\langle \varphi, v \rangle$ is the only pair of $\Gamma_\Delta^E(S)$ involving φ. Hence $\langle \varphi, v \rangle$ is in $\Sigma_\Delta(S)$, and so $E \subseteq \Sigma_\Delta(S)$ implying that $E \preceq_k \Sigma_\Delta(S)$.

2. Let v_1 and v_2 denote respectively the valuations defined by $\Sigma_\Delta(S_1)$ and $\Sigma_\Delta(S_2)$, and let $\langle \varphi, v \rangle$ be in $\Sigma_\Delta(S_1)$. If φ occurs in E, then the previous point shows that $\langle \varphi, v \rangle$ also belongs to $\Sigma_\Delta(S_2)$, and so $v_1(\varphi) = v_2(\varphi)$. Assuming now that φ does not occur in E implies that rules whose head is φ or $\neg\varphi$ have generated pairs involving φ. We consider the following distinct cases:

- Case 1: $v_1(\varphi) = t$ or $v_1(\varphi) = f$. Since in S_1, $v_\oplus(\varphi) = t$ (respectively $v_\oplus(\varphi) = f$), $\Gamma_\Delta^E(S_1)$ contains $\langle \varphi, t \rangle$ (respectively $\langle \varphi, f \rangle$) and not $\langle \varphi, b \rangle$. Then, by Definition 4(1), $inst(E, R)$ contains a rule ρ such that $head(\rho) = \varphi$ (respectively $head(\rho) = \neg\varphi$) and $v_{S_1}(body(\rho)) = t$. Then, for every l in $body(\rho)$, if $l = \phi$, $v_{S_1}(\phi) = t$ and if $l = \neg\phi$, $v_{S1}(\phi) = f$. Since $S_1 \preceq_k S_2$, for every ϕ occurring in $body(\rho)$, either $v_{S_2}(\phi) = v_{S_1}(\phi)$ or $v_{S_2}(\phi) = b$. Hence, whatever the chosen implication ($\rightarrow$ or $\hookrightarrow$), $v_2(\varphi) = v_1(\varphi)$ or $v_2(\varphi) = b$ and so, $v_1(\varphi) \preceq_k v_2(\varphi)$.

- Case 2: $v_1(\varphi) = b$. Either $\langle \varphi, b \rangle$ is in $\Gamma_\Delta^E(S_1)$ or not.

  (i) If $\langle \varphi, b \rangle$ is in $\Gamma_\Delta^E(S_1)$, then $inst(E, R)$ contains ρ such that $head(\rho) = \varphi$ or $head(\rho) = \neg\varphi$, and $v_{S_1}(body(\rho)) = b$. Hence, for every l in $body(\rho)$, if $l = \phi$, $v_{S_1}(\phi)$ is t or b and if $l = \neg\phi$, $v_{S1}(\phi)$ is f or b, and at least one of these values is b. Since $S_1 \preceq_k S_2$, $v_{S_2}(\phi)$ is either t or b when $v_{S_1}(\phi)$ is t or b, and $v_{S_2}(\phi)$ is either f or b when $v_{S_1}(\phi)$ is f or b, and at least one of these values is b. Therefore $\langle \varphi, b \rangle$ belongs to $\Gamma_\Delta^E(S_2)$ and so, in S_2, $v_\oplus(\varphi) = b$. Thus, $v_1(\varphi) \preceq_k v_2(\varphi)$.

  (ii) If $\langle \varphi, b \rangle$ is not in $\Gamma_\Delta^E(S_1)$, then $\langle \varphi, t \rangle$ and $\langle \varphi, f \rangle$ both belong to $\Gamma_\Delta^E(S_1)$, which implies that $\langle \varphi, b \rangle$ is in $\Sigma_\Delta(S_1)$. Hence $inst(E, R)$ contains one rule whose head is φ and one rule whose head is $\neg\varphi$, and these rules apply for computing $\Gamma_\Delta^E(S_1)$ and for (i) above. Thus, in S_2 we have $v_\oplus(\varphi) = b$, showing that $v_1(\varphi) \preceq_k v_2(\varphi)$.

  We have shown that for every φ occurring $\Sigma_\Delta(S_1)$, $v_1(\varphi) \preceq_k v_2(\varphi)$. Now, for every φ not occurring in $\Sigma_\Delta(S_1)$, $v_1(\varphi) = n$ which is the lowest truth value. Thus, $v_1(\varphi) \preceq_k v_2(\varphi)$ holds, showing that $\Sigma_\Delta(S_1) \preceq_k \Sigma_\Delta(S_2)$ holds as well. Therefore, the proof is complete. $\square$

B   Proof of Proposition 1

Proposition 1. *Given a database $\Delta = (E, R)$, Σ_Δ^* is a minimal (with respect to set inclusion) model of Δ.*

Proof. If Σ_Δ^* is not a model of Δ, then one rule ρ of $inst(E, R)$ is not valid in Σ_Δ^*. In this case, $head(\rho)$ is not valid, while the conjunct defined by $body(\rho)$ is valid. Denoting $head(\rho)$ by φ (respectively $\neg\varphi$), either $v_\Delta(\varphi) = \mathbf{n}$ or $v_\Delta(\varphi) = \mathbf{f}$ (respectively $v_\Delta(\varphi) = \mathbf{t}$). If $v_\Delta(\varphi) = \mathbf{n}$, since $body(\rho)$ is valid in Σ_Δ^*, we have $\Sigma_\Delta(\Sigma_\Delta^*) \neq \Sigma_\Delta^*$, which is not possible. Therefore, either $head(\rho) = \varphi$ and $v_\Delta(\varphi) = \mathbf{f}$, or $head(\rho) = \neg\varphi$ and $v_\Delta(\varphi) = \mathbf{t}$, which is not possible by Definition 4. Thus, Σ_Δ^* is a model of Δ.

To show that Σ_Δ^* is a minimal model, let σ be a nonempty subset of Σ_Δ^*, and assume that $S = \Sigma_\Delta^* \backslash \sigma$ is a model of Δ. Let k be the least integer such that $\Sigma^{k-1} \cap \sigma = \emptyset$ and $\Sigma^k \cap \sigma \neq \emptyset$. We notice that k exists such that $k > 0$ because, since S is a model of Δ, it holds that $E \subseteq S$ and so, since $\Sigma^0 = E$, we have $\Sigma^0 \cap \sigma = \emptyset$. Now, let $\langle \varphi, \mathbf{v} \rangle$ be in $\Sigma^k \cap \sigma$ but not in Σ^{k-1}. In this case, $v_S(\varphi) = \mathbf{n}$ and as above, there exists one rule ρ in $inst(E, R)$ such that $head(\rho)$ is either φ or $\neg\varphi$ and in Σ^{k-1}, $head(\rho)$ is not valid, while the conjunct defined by $body(\rho)$ is valid. Since $\Sigma^{k-1} \subseteq S$, $body(\rho)$ is valid in S while $head(\rho)$ is not. This is a contradiction and so, the proof is complete. $\qquad\square$

C   Proof of Proposition 2

Proposition 2. *Let $\Delta = (E, R)$ be such that for every rule ρ in R, $head(\rho)$ is a positive literal. For all minimal models M_1 and M_2 of Δ, the following holds:*
(i) $\mathsf{V}(M_1) = \mathsf{V}(M_2)$ and (ii) $\mathsf{F}(M_1) = \mathsf{F}(M_2)$.

Proof. The proposition is a consequence of Lemma 2 shown next. $\qquad\square$

Lemma 2. *Let $\Delta = (E, R)$ be such that for every rule ρ in R, $head(\rho)$ is a positive literal. For every minimal model M of Δ, the following holds:*

1. $\mathsf{F}(\Sigma_\Delta^*) = \mathsf{F}(M)$.
2. $\mathsf{V}(\Sigma_\Delta^*) = \mathsf{V}(M)$.

Proof. 1. As computing Σ_Δ^* starts from E and generates no other false facts, $\mathsf{F}(\Sigma_\Delta^*) = \mathsf{F}(E)$. Since $\mathsf{F}(E) \subseteq \mathsf{F}(M)$, we obtain $\mathsf{F}(\Sigma_\Delta^*) \subseteq \mathsf{F}(M)$.
   Assuming that $\mathsf{F}(M) \not\subseteq \mathsf{F}(\Sigma_\Delta^*)$, let φ be in $\mathsf{F}(M) \backslash \mathsf{F}(\Sigma_\Delta^*)$. Denoting by M' the set $M \backslash \{\langle \varphi, \mathbf{f} \rangle\}$, we show that M' is a model of Δ and thus that we obtain a contradiction since M is assumed to be minimal. To show that M' is a model of Δ, we first note that $E \subseteq M'$ holds because so does $E \subseteq M$ and $\langle \varphi, \mathbf{f} \rangle$ is not E. Thus, assuming that M' is *not* a model of Δ entails that there exists ρ in $inst(E, R)$ that is not valid in M'. Hence, independently from the chosen implication $\rightarrow$ or $\hookrightarrow$, $body(\rho)$ is valid in M' whereas $head(\rho)$ is not. However, since $body(\rho)$ is valid in M', φ does not occur in $body(\rho)$, implying that $body(\rho)$ is also valid in M. Hence, $head(\rho)$ must be valid in M and so, $head(\rho) = \varphi$. This is a contradiction with the fact that φ is assumed to be in $\mathsf{F}(M)$.
   2. We prove that $\mathsf{V}(\Sigma_\Delta^*) \subseteq \mathsf{V}(M)$ by induction on k. Indeed, $\mathsf{V}(\Sigma^0) \subseteq \mathsf{V}(M)$ holds because $E = \Sigma^0$. Then, for $k > 0$, assume that $\mathsf{V}(\Sigma^{k-1}) \subseteq \mathsf{V}(M)$ and

let φ be in $\mathsf{V}(\Sigma^k)\backslash\mathsf{V}(M)$. Since $\mathsf{V}(E) \subseteq \mathsf{V}(M)$, φ is not in $\mathsf{V}(E)$, φ occurs in Σ^k due to a rule ρ. Thus, there exists ρ in $inst(E, R)$ such that $body(\rho)$ is valid in Σ^{k-1} and $head(\rho) = \varphi$. Since $\mathsf{V}(\Sigma^{k-1}) \subseteq \mathsf{V}(M)$, $body(\rho)$ is valid in M and as ρ must be valid in M, so is φ. We thus obtain a contradiction with the fact that φ is assumed not to be in $\mathsf{V}(M)$. Hence $\mathsf{V}(\Sigma^k) \subseteq \mathsf{V}(M)$ and thus, $\mathsf{V}(\Sigma_\Delta^*) \subseteq \mathsf{V}(M)$.

Now let $M' = \{\langle\varphi,\mathsf{v}\rangle \in M \mid \varphi \in \mathsf{V}(\Sigma_\Delta^*)\} \cup \{\langle\varphi,\mathsf{f}\rangle \in M \mid \varphi \in \mathsf{F}(\Sigma_\Delta^*)\}$. We notice that $M' \subseteq M$, and since $\mathsf{V}(\Sigma_\Delta^*) \subseteq \mathsf{V}(M)$ and $\mathsf{F}(\Sigma_\Delta^*) = \mathsf{F}(M)$, we have $\mathsf{V}(\Sigma_\Delta^*) = \mathsf{V}(M')$ and $\mathsf{F}(\Sigma_\Delta^*) = \mathsf{F}(M')$. We show that M' is a model of Δ and thus that $M' = M$ since M is assumed to be minimal. To show that M' is a model of Δ, we first prove that $E \subseteq M'$. Indeed, as every $\langle\varphi,\mathsf{v}\rangle$ in E is also in Σ_Δ^* and in M, we have the following:

- If $\mathsf{v} = \mathsf{f}$ then φ is in $\mathsf{F}(E)$ thus in $\mathsf{F}(\Sigma_\Delta^*)$. In this case, $\langle\varphi,\mathsf{v}\rangle$ is in M'.
- Otherwise, $\mathsf{v} = \mathsf{t}$ or $\mathsf{v} = \mathsf{b}$, that is φ is in $\mathsf{V}(E)$. Thus φ is in $\mathsf{V}(\Sigma_\Delta^*)$ and as $\langle\varphi,\mathsf{v}\rangle$ is in M, $\langle\varphi,\mathsf{v}\rangle$ is also in M'.

Every rule ρ in $inst(E, R)$ is valid in M', because if $body(\rho)$ is valid in M' then $body(\rho)$ is also valid in Σ_Δ^* and so, $head(\rho)$ is in $\mathsf{V}(\Sigma_\Delta^*)$. Thus $head(\rho)$ is in $\mathsf{V}(M')$. Hence, $M' = M$, showing that $\mathsf{V}(\Sigma_\Delta^*) = \mathsf{V}(M)$. □

References

1. Afrati, F.N., Kolaitis, P.G.: Repair checking in inconsistent databases: algorithms and complexity. In: Fagin, R. (ed.) Proceedings of the 12th International Conference on Database Theory, ICDT 2009, pp. 31–41 (2009)
2. Arieli, O., Avron, A.: The value of the four values. Artif. Intell. **102**(1), 97–141 (1998)
3. Belnap, N.D.: A useful four-valued logic. In: Dunn, J.M., Epstein, G. (eds.) Modern Uses of Multiple-Valued Logic. EPIS, vol. 2, pp. 5–37. Springer, Dordrecht (1977). https://doi.org/10.1007/978-94-010-1161-7_2
4. Bergman, M.: The open world assumption: elephant in the room. In: AI3:::Adaptative Information, pp. 1–11 (2009). www.mkbergman.com/852/the-open-world-assumption-elephant-in-the-room/
5. Bidoit, N.: Negation in rule-based database languages: a survey. Theor. Comput. Sci. **78**(1), 3–83 (1991)
6. Ceri, S., Gottlob, G., Tanca, L.: Logic Programming and Databases. Springer, Heidelberg (1990). https://doi.org/10.1007/978-3-642-83952-8
7. Fitting, M.C.: Bilattices and the semantics of logic programming. J. Log. Program. **11**, 91–116 (1991)
8. Gelfond, M., Lifschitz, V.: The stable model semantics for logic programming. In: Proceedings of the International Conference and Symposium on Logic Programming, pp. 1070–1080 (1988)
9. Greco, G., Greco, S., Zumpano, E.: A logical framework for querying and repairing inconsistent databases. IEEE Trans. Knowl. Data Eng. **15**(6), 1389–1408 (2003)
10. Grant, J., Hunter, A.: Analysing inconsistent first-order knowledgebases. Artif. Intell. **172**(8–9), 1064–1093 (2008)
11. Greco, S., Molinaro, C., Trubitsyna, I.: Computing approximate query answers over inconsistent knowledge bases. In: Proceedings of the International Joint Conference on Artificial Intelligence, IJCAI, pp. 1838–1846 (2018)

12. Hazen, A.P., Pelletier, F.J.: K3, L3, LP, RM3, A3, FDE: how to make many-valued logics work for you. CoRR, abs/1711.05816 (2017)
13. Loyer, Y., Spyratos, N., Stamate, D.: Hypothesis-based semantics of logic programs in multivalued logics. ACM Trans. Comput. Log. 5(3), 508–527 (2004)
14. Reiter, R.: On closed world data bases. In: Logic and Data Bases, pp. 55–76 (1977)
15. Thimm, M.: On the expressivity of inconsistency measures (extended abstract). In: Proceedings of the International Joint Conference on Artificial Intelligence, IJCAI, pp. 5070–5074 (2017)
16. Tsoukiàs, A.: A first-order, four-valued, weakly paraconsistent logic and its relation with rough sets semantics. Found. Comput. Decis. Sci. 12, 85–108 (2002)

Query Driven Entity Resolution in Data Lakes

Giorgos Alexiou[1,2](✉) and George Papastefanatos[2](✉)

[1] School of Electrical and Computer Engineering, National Technical University of Athens,
Athens, Greece
[2] Information Management Systems Institute, ATHENA Research Center, Marousi, Greece
{galexiou,gpapas}@athenarc.gr

Abstract. Entity Resolution (ER) constitutes a core task for data integration which aims at matching different representations of entities coming from various sources. Due to its quadratic complexity, it typically scales to large datasets through approximate, i.e., blocking methods: similar entities are clustered into blocks and pair-wise comparisons are executed only between co-occurring entities, at the cost of some missed matches. In traditional settings, it is a part of the data integration process, i.e., a preprocessing step prior to making "clean" data available to analysis. With the increasing demand of real-time analytical applications, recent research has begun to consider new approaches for integrating Entity Resolution with Query Processing. In this work, we explore the problem of query driven Entity Resolution and we propose a method for efficiently applying blocking and meta-blocking techniques during query processing. The aim of our approach is to effectively and efficiently answer SQL-like queries issued on top of dirty data. The experimental evaluation of the proposed solution demonstrates its significant advantages over the other techniques for the given problem settings.

Keywords: Entity Resolution · Entity matching · Data lakes

1 Introduction

Entity Resolution (ER) is well studied problem in the data and web management communities, whose goal is to identify and match different representations of the same real-world entity. In traditional settings, it is a part of the data integration process, i.e., a preprocessing step prior to making "clean" data available to analysis. It is considered an expensive process as it requires the comparison of all entities from one data source with all other entities from the other sources. Hence, traditional approaches are often inexpedient for many modern query-driven applications that need to analyze only a small subset of the entire dataset and produce quick results from the data.

With the increasing demand of real-time analysis over heterogenous data sources, recent research has begun to consider new approaches for integrating Entity Resolution with Query Processing. A common setting for such scenarios is a data lake, where multiple heterogenous sources are aggregated and made directly available for analysis, avoiding the burdensome ETL tasks of a data warehouse. Data lakes usually contain duplicate entities from multiple sources which need to be resolved before enabling further

© Springer Nature Switzerland AG 2020
G. Flouris et al. (Eds.): ISIP 2019, CCIS 1197, pp. 117–130, 2020.
https://doi.org/10.1007/978-3-030-44900-1_8

analysis. For example, as shown in Fig. 1 let us consider a user who wishes to perform a comparative analysis on a data lake with scholarly data (e.g. publications) collected from three different sources s_1, s_2 and s_3 (for simplicity we assume that all exhibit similar relational schemas) containing highly overlapping publications. $\{P_1, P_2\}$, $\{P_3, P_4, P_5\}$ and $\{P_6, P_7\}$ are sets of the matching publications, coming however from different sources and thus exhibiting slight differences in the description of their attributes, e.g. authors' or venue names are abbreviated, some entities have missing years etc. The user would like to explore the contents of the three sources, and thus requests all papers, published in a specific conference (e.g., VLDB).

Source	PID	Title	Author	Venue	Year
S1	P1	Towards efficient Entity Resolution	Perry Scope	VLDB	2003
S2	P2	Towards efficient E.R.	Perry Scope	Very Large Data Bases	
S1	P3	Entity Resolution on web data	Allie Grater, John Doe	ACM SIGMOD	2016
S2	P4	E.R. on web data	A. Grater, J. Doe	SIGMOD Conference	
S3	P5	Entity-resolution on web data	A. Grater, John D.	Proc of ACM SIGMOD	2016
S1	P6	Entity-Resolution for scholarly data	Perry Scope, Emma Grate	VLDB	2015
S2	P7	E.R. for scholarly data	P. Scope, E. Grate	Very Large Data Bases	2015

Select * FROM pubs Where venue ="VLDB"

↓

PID	Title	Author	Venue	Year
P1	Towards efficient Entity Resolution	Perry Scope	VLDB	2003
P2	Towards efficient E.R.	Perry Scope	Very Large Data Bases	
P6	Entity-Resolution for scholarly data	Perry Scope, Emma Grate	VLDB	2015
P7	E.R. for scholarly data	P. Scope, E. Grate	Very Large Data Bases	2015

Fig. 1. Example of a query over 3 sources of scholarly data

In traditional settings, the query results will only contain $\{P_1, P_6\}$, which are exact matches for the condition *venue* = *'VLDB'*, excluding $\{P_2, P_7\}$, which although pertain to the same publications, their venue is differently described. To include in the results more entities, which could be matching with the ones retrieved by the original query, the user, following a rather naïve approach, will identify all possible alternative venue descriptions, perform a series of queries in the data lake, compare the results across the three sources, identify matching entities and finally group them based on a set of common characteristics (e.g., title, authors and venue). If the condition involves more attributes, e.g. the user wishes to filter based on the name of an author, the space of possible queries and follow-up tasks increase. Even in a more ETL-like scenario, the user would need to perform a preprocessing ER step to match all entities across the three sources, group them and construct a set of matching entities, in which all attributes are normalized, cleaned and semantically aligned, prior making it available for querying.

On the other hand, we would like to offer users the ability to perform all these time-consuming and error-prone operations directly on the raw data, during the same single query. In the example of Fig. 1, the entities P2 and P7 matching to the ones originally retrieved by the query will be also fetched and presented to the user. In this paper, we consider the problem of efficiently answering SQL queries on top of data lakes. Our goal is to minimize the effort for preprocessing and construction of a clean set and enable the user to operate directly on the raw heterogeneous data, i.e. the query is performed directly on the data lake avoiding the costly preprocessing step. Our main contribution is the *BlockJoin operator,* a technique that seamlessly integrates Entity Resolution in query execution. Our method takes an SQL query and employs blocking and meta-blocking techniques in order to enrich the query results with candidate matches from the underlying data lake. These techniques can dramatically reduce the overall number of the final comparisons (e.g., redundant comparisons), while maintaining the original number of matching ones; thus, improving the overall performance. The enriched results are then resolved for duplicates using established entity matching techniques, grouped and presented to the user. Such operations will enable users to perform more complex analytics directly on heterogenous sources avoiding the tedious preprocessing steps of batch Entity Resolution.

Outline. In Sect. 2 we present basic terms for our approach; in Sect. 3 we present the main concept of our approach for evaluating entity resolution task during the query evaluation, and in Sect. 4 we present the experimental evaluation of this approach. In Sect. 5 we provide related work, whereas in Sect. 6 we conclude the paper and provide insights for future directions.

2 Basic Concepts

In this section, we provide some preliminary concepts related to the entity resolution process, which are necessary for the presentation and experimentation of our approach in Sects. 3 and 4 respectively.

Data Lake. A set of heterogeneous entity collections describing overlapping real-world objects in different structures, formats and quality.

Entity Resolution (ER). ER is the task that identifies and aggregates the different entities/profiles that actually describe the same real-world object. The basic concept of ER is the entity e, alternatively called profile. As entity, we consider a set of name-value pairs that is associated with a unique id and describes a real-world object (i.e. a publication). A set of entities is called entity collection E. Two entities, e_i and e_j, considered to be duplicates, are notated with $e_i \equiv e_j$. This task is also called Deduplication [1, 2] in the context of homogeneous data collections.

Blocking. To scale ER, blocking restricts the executed comparisons to similar entities. It groups them into clusters, called blocks, based on the similarity of their attributes (e.g., tokens, n-grams, etc.) and performs pair-wise comparisons only between the entities of each block b_i. A set of blocks is called block collection B. Its size |B| denotes the number of blocks it contains, while its cardinality denotes the total number of comparisons it involves: $||B|| = \sum_{bi \in B} ||bi||$, where $||b_i||$ is the cardinality, i.e., number of comparisons in b_i.

Blocking methods [2–4] are widely used in ER. The main blocking methods for structured data have been summarized in the recent survey by Christen [2]. In this work, we incorporate the Standard Blocking technique (StBl) [5], although other techniques can also be used. It represents every entity by one or more keys and creates blocks on their equality; i.e., every block corresponds to a specific key and contains all entities that have it in their representation. For the key selection, we employ schema-agnostic blocking techniques [5], which are preferred for their increased effectiveness in terms of recall. A block is formed by each token found in all attributes of publications; this block contains the publications having this token in any of their attributes.

Blocking methods rely on redundancy; a blocking key places multiple entities in the same block, which results in many redundant (i.e., non-matching) and superfluous (i.e., existing in multiple blocks) comparisons. This way high recall is achieved at the cost of more comparisons (i.e. lower precision). This effect is partially improved by coarse-grained meta-blocking techniques that discard entire blocks either a-priori or during the resolution process.

Meta-blocking. The goal of meta-blocking is to restructure a given block collection B into a new one that contains significantly fewer redundant and superfluous comparisons, while maintaining the original number of matching ones [6]. The quality of a block collection B is measured in terms of two competing criteria: efficiency and effectiveness. The former is directly related to the total number of comparisons $\|B\|$ contains. The effectiveness of B depends on the cardinality of the set $D(B)$ of detectable matches (i.e., pairs of duplicate entities compared in at least one block). There is a clear trade-off between the effectiveness and the efficiency of B: the more comparisons are executed (i.e., higher $\|B\|$), the higher its effectiveness gets (i.e., higher $|D(B)|$), but the lower its efficiency is, and vice versa. Successful block collections achieve a good balance between these two competing objectives. Block processing methods are divided into two main categories according to the granularity of their functionality [4]:

1. *Block-refinement methods*, which operate at the coarse level of individual blocks, and
2. *Comparison-refinement methods*, which operate at the finer level of individual comparisons.

In this work we consider only the first ones because while they exhibit limited accuracy when discarding comparisons, they consume minimal resources, as they typically involve very low space and time complexity, which especially in the context of a query-driven ER are very crucial.

Query Entity Blocking. An entity collection in the context of QER is formed by the entities returned from a select query and is denoted as $QE \subseteq E$.

In QER, the block b_i is formed based on the entities retrieved by the select query and is denoted as Qb_i where $Qb_i \subseteq QE$. The size of Qb_i, i.e., the number of entities it contains, is denoted by $|Qb_i|$. The cardinality of Qb_i is denoted by $\|Qb_i\|$ and represents the total number of comparisons it contains: $\|Q_{b_i}\| = |Qb_i| \cdot (|Qb_i| - 1)/2$.

Entity Matching. We consider entity matching as an orthogonal task to blocking [1, 2, 4, 5]. That is, we assume that two duplicates are detected in B as long as they cooccur in at least one of its blocks. Provided that the vast majority of duplicate entities are co-occurring, the performance of ER depends on the accuracy of the method used for entity comparison.

2.1 Evaluation Measures

The effectiveness of ER, employed also in our QER problem, is assessed using the following measures:

Pairs Completeness (PC) estimates the recall of B, i.e., the portion of duplicates from the input entity collection(s) that cooccur in at least one block. More formally, PC = |D(B)|/|D(E)|. PC is defined in [0, 1], with higher values showing higher recall.

Pairs Quality (PQ) estimates the precision of B, i.e., the portion of its comparisons that correspond to duplicate entity profiles. More formally, PQ = |D(B)|/||B||. PQ takes values in the interval [0, 1], with higher ones indicating higher precision

Reduction Ratio (RR) estimates the portion of comparisons that not executed due to blocking in relation to the naive, brute force approach. Formally, it is defined as: RR(B, E) = 1 − ||B||/||E||, where ||E|| denotes the computational cost of the naive approach, i.e., ||E|| = |E| · (|E| − 1)/2. RR takes values in the interval [0, 1], with higher values indicating higher efficiency.

Overhead Time (OTime) is the total time required by the blocking method to cluster the given entities into blocks, i.e., the time between receiving the original query entity collection QE and returning a set of blocks as output and also the time for the meta-blocking methods to reduce the number of the final blocks.

Resolution Time (RTime) is the time required for performing all pair-wise entity' comparisons with a specific entity matching technique. As such, we consider the Jaro-Winkler similarity of the tokens in all attribute values of the compared entities.

The goal of QER is to maximize all measures keeping the time measures low. That is, to maximize the number of detected duplicates (|D(B)|), while minimizing the cardinality of B (||B||). In practice, though, there is a clear trade-off between PC and PQ: the more comparisons are executed (higher ||B||), the more duplicates are detected (higher |D(B)|), thus increasing PC; given, though, that ||B|| increases quadratically for a linear increase in |D(B)|, PQ and RR are reduced [7, 8]. Therefore, our approach should aim for a balance between precision (PQ) and recall (PC) that minimizes the executed comparisons and ensures that most matching entities cooccur (i.e., high PC).

3 Query Driven ER Approach

The overview of our approach is shown in Fig. 2. We enrich the traditional query answering flow with operators which implement ER methods during the query execution. We assume simple SQL Select-Project queries (SP) which are performed on the underlying RDBMS. The initial blocking index (BI) is constructed offline from all the entities in our dataset. The first step (1) selects the entities from the database that satisfy user's query criteria. Then, we create a query blocking index (QBI) (2) from the retrieved entities. For example, following Token blocking techniques [2] we consider that a block is formed by each token found in all attributes of publications; this block contains the publications having this token in any of their attributes. In (3), we introduce the block-join operator that performs a join between the initial blocking index (BI) and the query blocking index (QBI) in order to retrieve all those "dirty" (i.e. containing duplicates) subsets which approximately (possibly containing false-positives but not the opposite) answer the user's initial query. Then, at step (4) we apply meta-blocking techniques for reducing the final number of blocks, and thus the comparisons to be performed for identifying the duplicates. Next (5), we perform the matching between the entities in the blocks and our query entities to resolve the duplicates, and (6) the resolved entities are grouped in clusters. After this step, the results can be further used by any subsequent operation (project, join, analytics, etc.) needed to answer the query. The output can also be used to update the initial blocking index with the resolved ones to allow us to speed-up next queries by having a "cleaner" initial set. Following, we provide the details of each step.

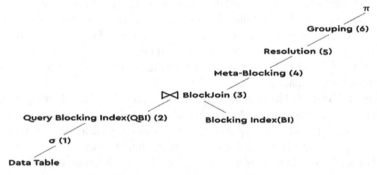

Fig. 2. ER-enriched Query-Plan

Blocking Index Initialization. The initial step of the proposed approach is to build the initial blocking index (BI). We follow an offline approach and build the BI only once, based on the schema – agnostic StBl. BI has a key-value data structure to store blocks. The keys are the tokens that derive from the StBl method (stop and common words are excluded) and the values are the IDs of the entities that correspond to the specific keys. Depending on its size this index can be stored directly in database or can be stored (i.e. serialized) in a data structure in order to load it directly in-memory (we follow the latter for efficiency reasons).

Query Block Index Creation. The query blocking index (QBI) is created for every single query that the user performs. It employs the same blocking technique with the BI but instead of indexing all the entities of the dataset it only indexes the entities that are retrieved from the user's query. The retrieved entities, in most of the cases, correspond to a small subset of the initial dataset thus the creation of this index is faster than the BI creation. In Fig. 3, a subset of the blocks in the QBI (query selects p_1 and p_6 entities) and in the overall BI is shown at the left side, corresponding to the example of Fig. 1. In the QBI, the block b_{vldb} contains p_1 and p_6, and $b_{towards}$ contains only p_1, whereas, in BI, b_{vldb} again contains p_1 and p_6, and $b_{towards}$ contains now p_1 and p_2.

BlockJoin Operator. The *BlockJoin* operator performs a hash-based join between the initial blocking index (BI) and the query blocking index (QBI), based on the keys of the QBI. The result is an enhanced QBI index with the same blocks being enriched with all the entities from the BI that correspond to these keys. In the output of Fig. 3 (the enhanced QBI), $b_{towards}$ contains now p_1 and p_2, b_{entity} contains p_3 and p_5, etc.

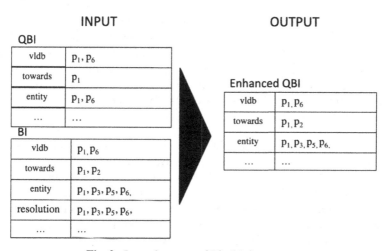

Fig. 3. Input & output of BlockJoin operator

Meta Blocking. In our approach the blocks that are derived from the BlockJoin operation contain many redundant and superfluous comparisons. Each token (i.e., blocking key) may be found in the attributes of many entities leading to many redundant comparisons. As explained in Sect. 2, we consider only *Block-refinement methods,* which operate only on the blocks' level because they consume minimal resources. More specifically we employ the *Block Purging* in conjunction with the *Block Filtering* methods, both explained below.

- *Block Purging* [9]. This method aims at cleaning the block processing list from oversized blocks. These are blocks that correspond to tokens of little discriminativeness, thus entailing a large number of comparisons while being unlikely to contribute nonredundant matches (i.e., duplicates whose profiles have no other token in common). Hence, they can be safely excluded from the block processing procedure, enhancing considerably the efficiency without any significant impact on the effectiveness.
- *Block Filtering* [9]: also relies on the idea that the larger a block is, the less likely it is to contain unique duplicates. Unlike Block Purging, though, it is applied independently to the blocks of every entity, assuming that each block has a different importance for every entity it contains. Block Filtering retains every entity in r% of its smallest blocks, with r typically set to 30–35% when applied after Block Purging.

Note that meta-blocking step is omitted in case the number of comparisons in the block index is small (i.e. below a threshold).

Resolution. The resolution step performs the comparisons between entities inside each block and resolves the matching ones. For each block in our block collection B, we take all its entities $|Qb_i|$ and we compare them with the entities derived from the user's select query $|QE|$. Thus the number of comparisons for each block will be: $||Q_{bi}|| = |Qb_i| \cdot (|Qb_i| - 1)/2$ and the total comparisons will be $||QB|| = \sum_{Qbi \in B} ||Qbi||$. For the comparisons of the entities we also employ schema-agnostic configuration, meaning that we compare all the values of the attributes of an entity e_i with the corresponding ones of an entity e_j. In this work we use the Jaro-Winker similarity function for the actual comparison of the values but any distance or similarity function could be used instead. In order to further improve our performance, at the time of the resolution we create a unique hashed ID of every comparison (based on the two entity ids plus a random string) to avoid repeated (i.e. redundant) comparisons. Those hashes are stored in a data-structure that offers constant-time access and are checked before each comparison (i.e. if the hash exists the comparison is omitted).

Grouping. Along with the step of resolution we employ a disjoint-set data structure (i.e. Union-Find) in order to store and group the resolved matched entities. By this way we manage to keep track of the duplicates by partitioning them into a number of disjoint (i.e. non-overlapping) subsets. We use this kind of data structure because it provides near-constant-time operations (bounded by the inverse Ackermann function). The disjoint-set data structure we employ is a forest; each tree corresponds to a set of matching entities, and each node of the tree corresponds to an entity having an entity ID, a parent entity, and a rank value. If an entity's parent element points to no other entity, then that entity is the root of a tree and is the representative member of its set. A set may consist of only a single entity (i.e. entity with no duplicates). However, if the entity has a parent, the entity is part of the tree identified by following the chain of parents upwards until the representative one is reached at the root of the tree.

The input of the grouping method is the two resolved entities belonging to a matching comparison processed during the resolution phase. The output of this step is a key-value data structure that has the root entity as key and its duplicates as values.

After this step, the results can be further used by any subsequent operation (project, join, analytics, etc.) needed to answer the query. The output (i.e. merged entities) can also be used to update the initial blocking index (BI) or the database with the resolved ones to allow us to speed-up next queries by having a "cleaner" initial set.

4   Experimental Evaluation

We implemented our approach[1] and experiments in Java version 8 and all source data and blocking index are stored in PostgreSQL version 11. The experiments were performed on a Mac computer with Intel i7 (2.2 GHz) and 16 GB of RAM DDR4, running on

---

[1] The source code and the queries are available in https://github.com/galexiou/isip2019.

macOS 10.14.5. All measurements were repeated 10 times and the average values are reported.

We have experimented with a real-world homogeneous dataset, employed in literature [6] for similar ER tasks. The *DBLP-Scholar* dataset [19] is an established dataset that has been widely used in literature [19–21]. It contains bibliographic 66879 records with 5347 matches from DBLP and Google Scholar (ground-truth is provided); the size of the cartesian product is approximately 2.2B entity pairs. We used a fixed blocking (StBl) and meta-blocking strategy for all experiments to guaranty equal effectiveness and efficiency of the workflow steps.

Regarding the meta-blocking techniques, Block Purging is parameter-free and Blocking Filtering involves a single parameter – the portion r of the retained blocks per entity. We set to $r = 0.30$, which retains every entity in around 1/3 of its most important (i.e., smallest) blocks. This value has been verified to reduce the total cardinality of blocks $\|B\|$.

To investigate the performance of our approach based on the measures described in 2.1, we used 10 (*select **) SP queries (Q1-Q10) with random sampling based on the modulo of the entities' IDs, ranging from 5% to 50% selectivity with 5% step. We evaluate our results on the meta-blocking output of the final comparisons and compare them with a *naïve approach*. The naïve approach is to consider that no blocking\meta-blocking operations are performed during query time and thus all entities retrieved from the select query are compared against all entities in the underlying dataset.

4.1 Experimental Results

The outcomes of our experiments with respect to performance are presented in Table 1. Starting with the number of comparisons of the naïve ($\|B\|$) approach per query we can observe that the number of the returned entities $|QE|$ yields a quadratic number of comparisons which obviously does not scale in voluminous datasets. On the other hand, with the application of blocking and meta-blocking techniques we managed to increase the performance by three orders of magnitude. It is also interesting to estimate the portion of saved comparisons in relation to the naïve approach (i.e., RR). We can do so by comparing $\|B\|$ with $\|QB\|$. In our case the RR remains stable as the selectivity grows and indicates the save of 99% of the possible comparisons.

Another interesting aspect is that our approach scales in a linear fashion to the increased size of the entities per query. Given that $Q10$ is almost 10 times larger than $Q1$, a linear increase in $\|QB\|$ requires that the blocks $|QB|$ for $Q10$ involve one order of magnitude more comparisons than those for $Q1$. This condition is satisfied in our experiments.

In Fig. 4 we examine the time requirements of our approach. It shows overhead time (*OTime*), resolution time (*RTime*) and the total number of comparisons performed by each per query. Starting with *OTime*, we can observe that it scales in a sub-linear fashion as the selectivity grows. This is an indication that the fact that we are using a schema-agnostic approach to create the blocks (i.e. using all the available attributes of the entities) does not seem to affect our blocking and meta-blocking techniques which seem to scale nicely. Regarding now the *RTime* (i.e. e time required for performing the pairwise comparisons in the resulting blocks), we can also see that it scales in a more

Table 1. |QE| stands for the number of entities returned from the query, |D(QE)| for the number of duplicate pairs in |QE|, |QB| for the number of comparisons after the meta-blocking step, ||QB|| for the number of comparisons executed, ||B|| for the number of comparisons of the naïve approach, RR for Reduction Ratio.

Queries	|QE|	|D(QE)|	|QB|	||QB||	||B||	RR
Q1	3343	502	9824	131764	111754809	0,99
Q2	6687	957	14903	222481	223576497	0,99
Q3	10133	1559	18675	304357	338809014	0,99
Q4	13375	1853	21540	366675	447219873	0,99
Q5	16719	2403	23749	431259	559041561	0,99
Q6	20266	2707	25905	492882	677651467	0,99
Q7	23466	3071	27478	542487	784657867	0,99
Q8	26751	3356	28879	586225	894506625	0,99
Q9	30125	3751	30060	627604	1007331498	0,99
Q10	33439	4034	31046	666703	1118150001	0,99

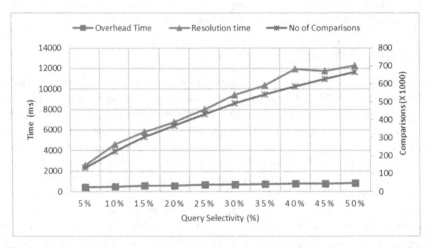

Fig. 4. QER overhead, resolution time and #Comparisons performed per query selectivity

linear fashion as the selectivity grows. *RTime* is greatly affected by the *OTime* as the blocking and meta-blocking methods save a considerable amount of comparisons which allow the actual resolution to scale. In our experiments, we attempted to measure the *RTime* of the *naïve approach;* however, for queries over 5% selectivity it was taking more than six hours. Thus, we can safely conclude that QER scales much better than the *naïve approach* and it also seems to be able to scale in voluminous datasets in acceptable query times.

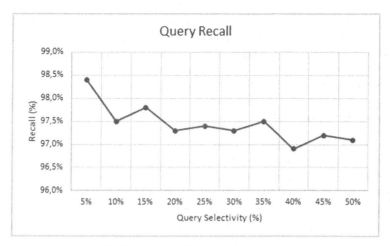

Fig. 5. QER recall per query selectivity

The outcomes of our experiments with respect to recall (*PC*) are presented in Fig. 5. The chart shows the percentage of the retrieved duplicates (Y axis) (based on the ground-truth) for each step of the query selectivity (X axis). We observe that QER maintains a robust recall (*PC*) that exceeds 97% (up to 98.5%) in all steps of the selectivity. This means that the initial set, in our case the entities retrieved by the query (*QE*), have a high number of recall and maintain it while the selectivity grows. That is to say that the recall seems to be selectivity independent. The schema-agnostic configuration [5] of our blocking process contributes a lot to this direction and enables us to achieve high and robust effectiveness.

5 Related Work

Entity Resolution. Given its importance, Entity Resolution (ER) has been studied thoroughly from the database community [10, 11]. Due to their quadratic complexity, existing ER approaches typically scale to large datasets through blocking methods which in principal compare only similar entities. Unlike the exhaustive ER techniques, blocking offers an approximate solution, sacrificing some recall in order to enhance precision. Primarily, the existing methods pertain to structured data, which abide by a specific schema with known semantics and qualitative characteristics for each attribute (Schema-based blocking) [2]. However, this approach is not applicable to Web Data, due to their highly heterogeneity. For that, blocking methods have been extended to function independent from the schema (schema-agnostic blocking), where every token from every value of every entity is treated as blocking key [5]. Although this approach successfully tackles the heterogeneity, it creates overlapping blocks resulting in unnecessary comparisons [5, 6]. Therefore, they must be avoided. This is achieved by block processing techniques that are appropriate for Web Data as they successfully tackle the heterogeneity as well as the great volume of the data. The most important method is the Meta-blocking [6, 9] which eliminates all the redundant comparisons along with the superfluous ones by examining the block-to-entity relationships. A notable work that tries to tackle the problem of ER in heterogeneous Web Data (e.g. Data Lake environments) is MinoanER [22]. It relies in Schema-agnostic techniques that consider the content and the neighbors of the entities in a progressive fashion, but it does it in a batch, offline processing of the entity collections. Instead, our approach operates online during query processing where the notion of progressive resolution is not applicable because we are focusing only on a subset of the whole dataset in each execution.

Query Driven Entity Resolution. Over the last years, a few methods for integrating Entity Resolution with Query Processing have been proposed with the aim to answer SQL-like queries performed over heterogenous data. Nevertheless, some of the existing methods are approximate solutions that are not designed for the larger class of SPJ (Select-Project-Join) queries [12], or even do not consider optimizing for other types of selection queries such as range queries or queries where the type of the condition attribute is not a string [13]. Other approaches are only considering the existence of probabilistic databases [14] in order to perform entity resolution techniques or are only answering a small class of topK and Iceberg queries [15]. One notable exception is the recent work presented in [16], which enables SQL query evaluation over dirty data that have been pre-processed and grouped together in corresponding blocks, based on specific blocking keys. This approach requires that (i) there exists a universal, well-defined schema underlying the data, and (ii) the optimal query plan is given by the user to the query engine. Unfortunately, none of these two requirements are satisfied in the case of heterogeneous Data Lakes. The proposed approach is designed to be able to scale in such environments with the use of minimum or none configuration (e.g. schema-alignment) by the end user.

Similarity Joins. A related approach to this specific problem is the Set Similarity Joins (SSJ) which compute all pairs of similar sets from two collections of sets. The recent survey by Fier et al. [17] though, which surveyed ten recent, distributed set similarity join algorithms had some interesting results: All algorithms in their tests failed to scale for at least one dataset and were sensitive to long sets, frequent set elements, low similarity thresholds, or a combination thereof. Moreover, some algorithms even failed to handle the small datasets that can easily be processed in a non-distributed setting.

6 Conclusions and Future Work

In this paper, we have proposed a Query Driven Entity Resolution approach applicable in heterogenous data storage settings, like the data lakes, where data is cleaned "on-the-fly" in the context of a query. We have developed a method to seamlessly integrate Entity Resolution techniques during query processing over heterogenous data. We have introduced a novel blockJoin operator to let us enable that kind of integration. Such operations will enable users to perform more complex analytics directly on heterogenous sources avoiding the tedious tasks of schema alignment and data deduplication.

This research opens several interesting directions for future investigation. First, our approach does not only target relational sources but non-relational ones (e.g., JSON files,

RDF stores) as long as they can be queried via an SQL interface; most commercial query engines offer SQL support over non-relational sources via proper functions. Next, while selection queries (as studied in this paper) are an important class of queries on their own, developing techniques for other types of queries (e.g., joins) is an interesting direction for future work. Another direction is developing a mechanism for efficient updates of the blocking index state for subsequent querying and finally the application our framework to distributed environments to enable big data processing. We also plan to incorporate a persistent merging process [18] in the steps of the current workflow that will also help us to better clean and update the original dataset and present the final results to the end-user.

Acknowledgements. This research is funded by the project VisualFacts (#1614) - 1st Call of the Hellenic Foundation for Research and Innovation Research Projects for the support of post-doctoral researchers.

References

1. Christen, P.: Data Matching. Data-Centric Systems and Applications. Springer, Heidelberg (2012). https://doi.org/10.1007/978-3-642-31164-2
2. Christen, P.: A survey of indexing techniques for scalable record linkage and deduplication. IEEE Trans. Knowl. Data Eng. **24**(9), 1537–1555 (2012)
3. Baxter, R., Christen, P., Churches, T.: A comparison of fast blocking methods for record linkage. In: Workshop on Data Cleaning, Record Linkage and Object Consolidation, pp. 25–27 (2003)
4. Papadakis, G., Ioannou, E., Palpanas, T., Niederee, C., Nejdl, W.: A blocking framework for entity resolution in highly heterogeneous information spaces. IEEE Trans. Knowl. Data Eng. **25**(12), 2665–2682 (2013)
5. Papadakis, G., Alexiou, G., Papastefanatos, G., Koutrika, G.: Schema-agnostic vs schema-based configurations for blocking methods on homogeneous data. Proc. VLDB Endow. **9**(4), 312–323 (2015)
6. Papadakis, G., Koutrika, G., Palpanas, T., Nejdl, W.: Meta-blocking: taking entity resolution to the next level. IEEE Trans. Knowl. Data Eng. **26**(8), 1946–1960 (2013)
7. Getoor, L., Machanavajjhala, A.: Entity resolution: theory, practice & open challenges. Proc. VLDB Endow. **5**(12), 2018–2019 (2012)
8. Getoor, L., Machanavajjhala, A.: Entity resolution for big data. In: KDD, p. 1527 (2013)
9. Papadakis, G., Papastefanatos, G., Palpanas, T., Koubarakis, M.: Scaling entity resolution to large, heterogeneous data with enhanced meta-blocking. In: EDBT, pp. 221–232 (2016)
10. Ipeirotis, P.G., Verykios, V.S., Elmagarmid, A.K.: Duplicate record detection: a survey. IEEE Trans. Knowl. Data Eng. **19**(1), 1–16 (2007)
11. Lenzerini, M.: Data integration: a theoretical perspective. In: Proceedings of the Twenty-First ACM SIGMOD-SIGACT-SIGART Symposium on Principles of Database Systems, pp. 233–246. ACM, June 2002
12. Altwaijry, H., Kalashnikov, D.V., Mehrotra, S.: Query-driven approach to entity resolution. Proc. VLDB Endow. **6**(14), 1846–1857 (2013)
13. Bhattacharya, I., Getoor, L.: Query-time entity resolution. J. Artif. Intell. Res. **30**, 621–657 (2007)
14. Ioannou, E., Nejdl, W., Niederée, C., Velegrakis, Y.: On-the-fly entity-aware query processing in the presence of linkage. Proc. VLDB Endow. **3**(1–2), 429–438 (2010)

15. Ioannou, E., Garofalakis, M.: Query analytics over probabilistic databases with unmerged duplicates. IEEE Trans. Knowl. Data Eng. **27**(8), 2245–2260 (2015)
16. Altwaijry, H., Mehrotra, S., Kalashnikov, D.V.: Query: a framework for integrating entity resolution with query processing. Proc. VLDB Endow. **9**(3), 120–131 (2015)
17. Fier, F., Augsten, N., Bouros, P., Leser, U., Freytag, J.C.: Set similarity joins on MapReduce: an experimental survey. Proc. VLDB Endow. **11**(10), 1110–1122 (2018)
18. Alexiou, G., Meimaris, M., Papastefanatos, G.: Enabling persistent identification of groups of duplicates in data aggregators. In: 2016 IEEE 32nd International Conference on Data Engineering Workshops (ICDEW). pp. 124–126. IEEE, May 2016
19. Köpcke, H., Thor, A., Rahm, E.: Evaluation of entity resolution approaches on real-world match problems. Proc. VLDB Endow. **3**(1–2), 484–493 (2010)
20. Kopcke, H., Thor, A., Rahm, E.: Learning-based approaches for matching web data entities. IEEE Internet Comput. **14**(4), 23–31 (2010)
21. Thor, A., Rahm, E.: MOMA-a mapping-based object matching system. In: CIDR, pp. 247–258, January 2007
22. Efthymiou, V., Stefanidis, K., Christophides, V.: Minoan ER: progressive entity resolution in the web of data. In: EDBT 2016, pp. 670–671 (2016)

Data Mining Applications

A Hybrid Recommender System
for Steam Games

Jin Gong, Yizhou Ye, and Kostas Stefanidis[✉]

Tampere University, Tampere, Finland
{jin.gong,yizhou.ye,konstantinos.stefanidis}@tuni.fi

Abstract. A recommender system can be considered as an information filtering system that seeks to predict the preference a user would have for a data item. It is commonly utilized in digital stores to recommend products to their users according to the users' previous purchases. This applies to Steam as well, a widely used digital distribution platform for games. The existing recommender system mainly suggests new games to a given user by calculating similarities between games they own and those that they do not. These similarities are based on predefined attributes (game genres). Additionally, the system is able to recommend games based on the game preferences of the user's friends. In this work, we target at creating an enhanced recommender system for Steam. The goal is to design a hybrid approach for producing suggestions that will utilize data, such as playing time, game price and game release date, in addition to the genres and the preferences of friends.

Keywords: Steam · User profile · Recommendation system · Collaborative filtering

1 Introduction

The Steam platform is the largest digital distribution platform for PC gaming nowadays. On 14 Jan 2019, Steam published its annual report based on the past 12 months, including data on stores, the Steam community, gameplay, Steamworks, and things behind the scenes. According to the report, Steam users have experienced explosive growth in 2018 [18]. Among them, the daily active users of the platform are up to 47 million, the monthly active users are 90 million, the highest number of online users is 18.5 million, and the monthly growth of users with valid purchases is 1.6 million. One of the key reasons Steam is growing so rapidly is the good search-ability in store, which was mentioned in the report as well. They are working on a new recommendation system driven by machine learning to find games that match the player's personal preferences. Although the algorithm is just part of the search-ability solution, they are also building more live and appreciation features and continually evaluating the overall design of the store.

On the other hand, the recommender system sometimes does not work as well as expected. One reason for this is the *Matthew Effect* [15], which means

© Springer Nature Switzerland AG 2020
G. Flouris et al. (Eds.): ISIP 2019, CCIS 1197, pp. 133–144, 2020.
https://doi.org/10.1007/978-3-030-44900-1_9

the rich gets richer and the poor gets poorer always appears in the social science field, can be also applied to the game market. Those games developed by big game developing companies receive more budget on advertisement so that they can be very popular. Popular games will appear on the top position on the store web pages and attract more users to buy including you and your friends. Meanwhile, games developed by small studios or individual developers are not that lucky. Those budget games without enough attention would easily disappear from users. What is even worse, if developers are unable to earn money from those games, they are very likely to break down which does harm to the whole game market. This is the reason why people need a recommender system, for suggesting niche games to users.

Our goal, in this paper, is to create a hybrid approach for producing suggestions that will utilize data, such as playing time, game price and game release date, in addition to the genres and preferences of friends used already. Our initial target is to analyze the data, which comes in three parts. The first part includes the user IDs, the games that they purchased and the hours they had spent playing the game[1]. The second part consists of the game titles combined with their prices and release dates[2], while the third part consists of the game names, game IDs and their genres. This dataset was manually crawled from the Steam API.

We convert all available data into numerical ratings ranging from 1 to 5. These ratings will then be used in calculating the Pearson correlation between the cases to determine the similarities. A rating of 1 equals to total similarity and -1 means that the entities being compared are total opposites to each other. For producing recommendations, the final rating of a game to any given user is the mean of several values, including user preference of this game genre, user preference of games similar to this game, similar users preference of this game, user preference of price and user preference of game released in that time zone. The proposed approach includes no assigned weights to the individual parts for computing the overall ratings, i.e., each aspect of the data would have an equal amount of impact on the final rating. Python is used as the programming language of the system.

We evaluate the accuracy of our results by separating the dataset into a test set and a training set. Specifically, we attempt to predict the values in the test set using the training set.

This paper is organized as follows. Section 2 discusses about the recommender system in use of the Steam platform and the recommender of other platforms. Section 3 introduces the dataset and analysis method. Section 4 shows the performance of the proposed approach, and Sect. 5 concludes the paper with a summary of our contributions.

---

[1] https://www.kaggle.com/tamber/steam-video-games/.
[2] https://www.kaggle.com/kingburrito666/over-13000-steam-games/.

2   Related Work

2.1   Recommender System of Steam

Valve Software[3] is keeping improve the recommending system of Steam platform annually. However, Valve has never gave out any detailed information about the algorithm of its recommender system. The system is like a black box to common users. Some analysts (e.g., Erik Johnson) have attempted to dive deeply into the mechanism behind Steam. Erik Johnson stopped using his personal Steam account and spent two months recording all the games he had played, viewed and commented since then, as well as the play-time, and he even stored the HTML pages of those games. During the two months, he viewed all the 672 games recommended by Steam and started to find the relationships among them.

According to this study, Eric Johnson found that the system could make better recommendations if it relied less on popularity and recency, and instead did a better job of surfacing titles based on quality and personal relevance factors. The challenge here is that popularity and recency are easy to quantify. Quality and relevance are more elusive. Furthermore, the most surprising omission in all these systems is the lack of collaborative filtering [10].

2.2   Other Works on Recommender Systems

The recommender system is not a unique feature of the game platform. Since the popularity of the Internet, various platforms have used their unique recommender systems to provide customized services to users.

In general, a recommender system aims at providing suggestions to users or groups of users by estimating their item preferences and recommending those items featuring the maximal predicted preference. Typically, depending on the type of the input data, i.e., user behavior, contextual information, item/user similarity, recommendation approaches are classified as content-based [19], collaborative filtering [20], knowledge-based [4], hybrid [2], or even social ones [22]. Nowadays, recommendations have more broad applications, beyond products, like links (friends) recommendations [28], query recommendations [6], health-related recommendations [23,24], open source software recommendations [11], diverse venue recommendations [7], recommendations for groups [14,16,17], sequential recommendations [3,25] or even recommendations for evolution measures [21,27].

Next, we take as an example a shopping experience to showcase how recommenders work. So, typically, recommender algorithms start by finding a set of customers who purchased and rated items overlap the user's purchased and rated items. The algorithm aggregates items from these similar customers, eliminate items the user has already purchased or rated, and recommends the remaining items to the user. In the case of item-to-item collaborative filtering, like in our work, the focus is on finding similar items, not similar customers. For each of the

---

[3] https://www.valvesoftware.com/en/.

user's purchased and rated items, the algorithm attempts to find similar items. It then aggregates similar items and recommends them (e.g., [13]). This is the method that many companies, like Amazon, are using. However, this method is more likely to be a *have-to* choice due to the lack of *friends* feature, so its recommendations rely on *items* and similar users. It would guess which product you may like according to your wish list, your previous purchase and the goods you searched for.

There are also shopping websites which allow users to add each other as *friends* and take what your friends bought into account. Some people may have similar experience like: *The shop recommends the item I just bought.* Here we are going to introduce a concept that is the cost of making mistakes. A bad recommendation will have a bad effect, but the question is how big the effect. The cost is very small in shopping field because when a user opens the website, he/she knows clearly what he/she needs. If they recommend him/her something, he/she does not need, this user will even never click on it. The game field is kind of the opposite. Many users open the game store without a specific game need. They just select a game for fun, no matter which one. It is very difficult to judge a game according to its description and several images. For example, based on the description and images, the user might think that he/she likes it, but after trying the game, he/she disliked it. During this period, the user needs to pay both money and time, leading to trust reduction of the user for the system.

This is totally different in another field, like YouTube. The top sector on YouTube web page is the recommendation of the video. Its algorithm is based on the channels you subscribed, the videos you previously viewed and videos you liked. They use two neural networks. The first one is the candidate generation network, it takes events from the user's YouTube activity history as input and retrieves a small subset (hundreds) of videos from a large corpus. The second one, the banking network, it accomplishes this task by assigning a score to each video according to the desired objective function, using a rich set of features describing the video and user. The highest scoring videos are presented to the user, ranked by their score [5]. For new user without any interactions with the system, the system will show the most trending videos based on the user's location. This mechanism is more like the Steam platform. It will classify users with many tags. For instance, if you have viewed many technology videos, you may gain a tag says *technology fans* and you may get as well other tags like *nature fans* or *fans of a pop star*, if you watch videos of that type. The steam platform is doing the same thing as well. If you played a lot of free games, they are highly likely to introduce other free game rather than those very expensive fee-paying ones to you.

3   The Dataset

To do deep analysis with Steam users' gameplay, a sufficient amount of data is needed. For our analysis, 100,000 users will be used. The open-source data company with educational datasets should be the best choice, namely, Kaggle.com.

Table 1. Dataset information.

Type	Size	Description
Users	11350	The number of users
Games	5155	The number of games
Games per user	17.62	The number of games owned by a user in average
Gaming time (in hours)	0.1–11754	The time a user spends on a game
Year of publish	2007–2017	The year the game published in

According to the Terms and Use of Kaggle.com, Steam's dataset can be downloaded for academic use, obeying any ethics issues. We found three different datasets from Kaggle. The first one [26] includes user IDs, the games that they purchased and the hours they had spent playing the game. The second dataset [12] includes the game titles combined with their prices and release dates. The third one only includes the game titles initially, however, for the hybrid recommendation system that we are aiming to create, we want to add the game genres into the mix.

A web crawler based on Python language was created to collect game genres from public game profiles. Steam URLs are of the form '/gameID/gameName', and we had no access to gameIDs in our data. After realized that using the Steam webpage would be far too inconvenient due to game name in the URL being inconsistent with the name provided in the data, the next attempt was made using a website containing information on all Steam games, SteamDB. Much to our dismay, only trials and errors are learned because this website actively blocks all crawling scripts, although we did manage to find the source that this website is using and ended up using that instead [1]. The gameIDs along with the game genres would then be added to the third dataset. However, many games (2030 cases) have missing genres due to several reasons: Some of them were old and thus removed from the current Steam store. Others could not be found because their names were spelled differently in the sources that we were comparing.

Table 1 shows the basic information about the dataset. If the whole dataset is changed into a user-game table, the known data (the user-game pair) only fills 0.34% of the whole table. The data itself is very sparse so any analysis based on the raw data is inefficient and inaccurate. Features should be extracted from the dataset for more research.

4   The Method

All available datasets are in a csv-format. Python libraries, like Numpy and Pandas, are used to organize the data. In the first dataset, there was an empty column that we removed. All the cases with missing genres are excluded.

Next, we transform the initial three datasets into four different tables.

Table 1: User-Game Table. We will first create the user-game table, which is a table containing users along with their ratings for the games based on the time they had spent playing the game. Our rating system is an interval from 1 to 5, and it was calculated by dividing the playing time into 5 equally parts [29]. For example, a given user would get a rating of 5 for a given game if the time the user had spent playing the game belonged in the top-20% out of all the users.

Table 2: User-Genre Table. We will then create the user-genre table by taking the average of all the ratings for games that belong in the same genre [8]. For example, let us assume that a user has game1 and game2. Game1 belongs to genre1 and genre2, game2 belongs to genre2. The rating this user would obtain for genre1 would be the rating they got for game1. Likewise, the score for genre2 would be the mean of game1 and game2. The formula can be generalized as follows:

$$score(user\_i, genre\_j) = \frac{\sum\limits_{g \in G_n} game_g\_rating}{n} \tag{1}$$

where n equals the number of games that $user_i$ has in $genre_j$, and G_n is the set of n games that $user_i$ has in $genre_j$.

Table 3: User-Price Table. We started by defining three price ranges:

- Free to Play games.
- Games that are below 20$, but not free.
- Games that are above or equal to 20$.

First, we calculate the number of games that a user has in each price range. The rating for that price range would then be the mean of all the ratings for games in that price range, e.g., if a given user had 4 games, game1 (free), game2 (<20$), game3 (≥20$) and game4 (≥20$), then the rating this user has for free games would be the rating for game1, the rating for games under 20$ would be the rating for game2 and the rating for games over 20$ would be the average of *rating_game3* and *rating_game4*.

Table 4: User-Release Year Table. It should be noted that the data we had only had release dates ranging from 2007 to 2017, which may cause some bias in our results, though Steam does remove older titles from the store regularly. They remain playable if the user had purchased them but cannot be bought from the store anymore. We categorized these release dates into three categories:

- Older than 2010.
- Games released between 2010 and 2015.
- Newer than 2015.

In this case, the rating system works exactly like in the previous user-price table, which is to say, the rating for a certain interval is the mean of all the ratings the user had given for games that belong in that interval.

Overall Aggregation. Pearson correlation is used to measure the similarities between the users and the similarities between the games. A score of 1 would equal total similarity and -1 would mean that the entities being compared are total opposite of each other. The Pearson correlation similarity of two users x, y is defined as:

$$simil(x, y) = \frac{\sum\limits_{i \in I_{xy}} (r_{x,i} - \bar{r}_x)(r_{y,i} - \bar{r}_y)}{\sqrt{\sum\limits_{i \in I_{xy}} (r_{x,i} - \bar{r}_x)^2} \sqrt{\sum\limits_{i \in I_{xy}} (r_{y,i} - \bar{r}_y)^2}} \tag{2}$$

where I_{xy} is the set of items rated by both user x and user y, $r_{a,b}$ the rating assigned to game b by user a, and $\bar{r}_a$ the mean of the ratings for user a.

After obtaining the similarity matrices, the system is able to produce recommendations of similar users for any given user. The same is true for games. This is accomplished using the prediction formula which is the same as what we use to calculate the correlation of two users. But the explanation should be the Pearson correlation similarity of two items x, y is defined as that, and where I_{xy} is the set of users give rate to both game x and game y.

Next, let us explain the system with a practical example: Bob does not own the game Dota2. Dota2 belongs to several genres: Action, Free to Play and Strategy. As is apparent from the genres, it is free. It was released in 2013. The final rating that Bob receives for Dota2 is the mean of five different ratings:

- The mean rating for Dota2 given by top-5 most similar users.
- The mean rating is given by Bob for top-5 most similar games.
- The rating Bob gave for games released between 2010 and 2015.
- The mean rating Bob gave for Action, Free to Play and Strategy-games.
- The rating Bob gave for games that are Free to Play.

5 Experiments

We evaluated the accuracy of our results by randomly selecting 10% of the data as the test set. The remaining 90% of the data would be used as the training set. We would randomly select one rating from a user and delete it. Subsequently, we would try to predict the deleted value using the training set. These predictions were then analyzed through MAE and NMAE [9]:

$$MAE = \frac{\sum\limits_{i \in N} |p_i - q_i|}{n} \tag{3}$$

$$NMAE = \frac{MAE}{R_{max} - R_{min}} \tag{4}$$

50 independent tests with random seeds have been done to select the different testing set and training set. The testing results were transformed into charts for better understanding (see, Figs. 1 and 2).

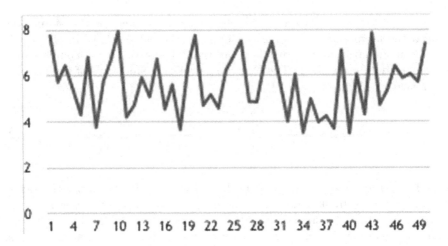

Fig. 1. MAE scores in 50 iterations.

It can be seen from the charts, the average of MAE is around 6 which stands for the average of prediction is about 2.45, and the average of NMAE is around 0.33 which means our model has an accuracy around 67% on predicting. Which means, every time the user is viewing a video game product, this system will automatically recommend five other video games according to the current game and the preference of the current user. Among the five recommended games, more than 2 of them (actually 2.45 out of 5) will meet the interest of the user. Meanwhile, if the system recommend a game to the user, it is 67% sure that the user will interested in this game or even purchase.

Another test is about how the accuracy change when user and game collaborative filtering, game genre, price and publishing year are considered one by one (see, Fig. 3). In this test, the system tries to predict the game that the user likes best or in other words, the games with longest playtime. We randomly select 10% of the data as the test set as well. The remaining 90% of the data would be used as the training set. For all the users in testing set, the playtime information is hidden, we can only know the names of the games they owned, but we do not know how long they spent on the games. The game which a user spent the longest time on is defined as his or hers favourite game. The accuracy is calculated as:

$$accuracy = \frac{N_c}{N},$$

where N_c represent the number of users that predicted correctly and N is the number of users contained in testing set.

If the system recommends games taking into account similarities between users, which in turn means similarities between the games the users own, we calculate similarities using the Pearson correlation (see, Sect. 4).

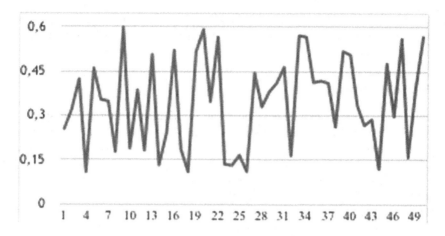

Fig. 2. NMAE scores in 50 tests.

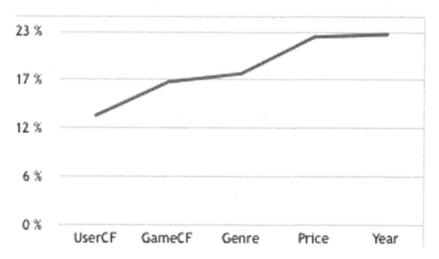

Fig. 3. The accuracy changes.

With this mechanism (UserFC), the average accuracy is about 13.5%. The second step is take both similar users and similar games into concern. For each game in the library, calculates the sum of the similarities between the game and all the other games owned by the user. The game with the highest sum score is seen as the game that should be recommended from GameCF. If the UserCF and GameCF give out different games to recommend, just choose the one more popular in all the other users. Then find out the favourite genre, price range and year range of the user, and recommend a game with highest score satisfying one feature above, we proceed as follows. The accuracy of the selected game is just the favourite game of the user going from 13.5% to 22.8%. Among the five factors, the price contribute the most and the publish year is the least important.

The result shows that even though from every aspect a game seems that it suits the user very well, but if the price is too expensive, the user will not choose the game.

6   Conclusion

A recommender system can be considered as an information filtering system that seeks to predict the preference a user would have for a data item. In this paper, we focus on the Steam, a widely used digital distribution platform for games. Specifically, we target at producing an enhanced recommender for Steam that uses data, such as playing time, game price and game release date, genres and users preferences for making suggestions.

Based on our first experimental results, we investigate that the big sparsity of our dataset appears to be an important reason for making the results not as good as expected. This is caused by two reasons. The first reason is the raw data itself that is too sparse and cannot be modified. The second one is data loss when different tables are joint together. For instance, a user has 10 games, but we only have information about 5 out of 10. Then, only those 5 games contribute to the analysis, which is still related to data sparsity.

Another reason is related to the fact, that on the particular domain, it is sometimes random if a user owns a suitable game or not. A game might be very suitable to the taste of one user, no matter what system you use to test, but the fact is that this user does not own the game. He or she might have already had this game on another platform or will meet the game in the future and then own it.

Among the 5 factors listed in the paper, a recommendation based on similar user and similar games seems more reliable which is so-called collaborative filtering. Meanwhile, the price of a game is a factor considered by most of the users tested. The age of the game and the genres seems to be not so important to many users. As we know, the genres are classified by tags contributed by other users and not supervised by Steam platform or game publisher. Misclassification sometimes happens on Steam. It could be a potential reason making genre factor not influential. To generalize, we opt in our future work to assign different weights to different factors, depending on their importance, instead of having all factors that equally effect on the final hybrid rating.

We also tried to generate rules from our results using the IBM's SPSS software tool. Using the Scikit library of Python, we were able to generate some rules, some even with very high accuracy, but the general issue was their low coverage (around 0,01% at best). Thus, we discarded these rules due to overfitting. We leave this as future work, and we will consider in our next steps, alternative ways for generating rules.

References

1. Official team fortress wiki. https://wiki.teamfortress.com/wiki/. Accessed 11 Apr 2019

2. Balabanovic, M., Shoham, Y.: Content-based, collaborative recommendation. Commun. ACM **40**(3), 66–72 (1997)
3. Borges, R., Stefanidis, K.: Enhancing long term fairness in recommendations with variational autoencoders. In: Proceedings of the 11th International Conference on Management of Digital EcoSystems, MEDES 2019 (2019)
4. Bridge, D.G., Göker, M.H., McGinty, L., Smyth, B.: Case-based recommender systems. Knowl. Eng. Rev. **20**(3), 315–320 (2005)
5. Covington, P., Adams, J., Sargin, E.: Deep neural networks for YouTube recommendations. In: Proceedings of the 10th ACM Conference on Recommender Systems, pp. 191–198. ACM (2016)
6. Eirinaki, M., Abraham, S., Polyzotis, N., Shaikh, N.: QueRIE: collaborative database exploration. IEEE Trans. Knowl. Data Eng. **26**(7), 1778–1790 (2014)
7. Ge, X., Chrysanthis, P.K., Pelechrinis, K.: MPG: not so random exploration of a city. In: MDM (2016)
8. Goldberg, D., Nichols, D., Oki, B.M., Terry, D.: Using collaborative filtering to weave an information tapestry. Commun. ACM **35**(12), 61–71 (1992)
9. Hu, Y., Koren, Y., Volinsky, C.: Collaborative filtering for implicit feedback datasets. In: 2008 Eighth IEEE International Conference on Data Mining, pp. 263–272. IEEE (2008)
10. Johnson, E.: A deep dive into steam's discovery queue. https://www.gamasutra.com/blogs/ErikJohnson/20190404/340061/A_Deep_Dive_Into_Steams_Discovery_Queue.php. Accessed 7 July 2019
11. Koskela, M., Simola, I., Stefanidis, K.: Open source software recommendations using github. In: Méndez, E., Crestani, F., Ribeiro, C., David, G., Lopes, J.C. (eds.) TPDL 2018. LNCS, vol. 11057, pp. 279–285. Springer, Cham (2018). https://doi.org/10.1007/978-3-030-00066-0_24
12. Larson, L.: Over 13,000 steam games. https://www.kaggle.com/kingburrito666/over-13000-steam-games/. Accessed 11 Apr 2019
13. Linden, G., Smith, B., York, J.: Amazon.com recommendations: item-to-item collaborative filtering. IEEE Internet Comput. **7**(1), 76–80 (2003)
14. Machado, L., Stefanidis, K.: Fair team recommendations for multidisciplinary projects. In: 2019 IEEE/WIC/ACM International Conference on Web Intelligence, WI 2019, Thessaloniki, Greece, 14–17 October 2019, pp. 293–297 (2019)
15. Merton, R.K.: The Matthew effect in science: the reward and communication systems of science are considered. Science **159**(3810), 56–63 (1968)
16. Ntoutsi, E., Stefanidis, K., Nørvåg, K., Kriegel, H.-P.: Fast group recommendations by applying user clustering. In: Atzeni, P., Cheung, D., Ram, S. (eds.) ER 2012. LNCS, vol. 7532, pp. 126–140. Springer, Heidelberg (2012). https://doi.org/10.1007/978-3-642-34002-4_10
17. Ntoutsi, E., Stefanidis, K., Rausch, K., Kriegel, H.: Strength lies in differences: diversifying friends for recommendations through subspace clustering. In: CIKM (2014)
18. O'Neill, M., Vaziripour, E., Wu, J., Zappala, D.: Condensing steam: distilling the diversity of gamer behavior. In: Proceedings of the 2016 Internet Measurement Conference, IMC 2016, pp. 81–95. ACM, New York (2016). https://doi.org/10.1145/2987443.2987489. http://doi.acm.org/10.1145/2987443.2987489
19. Pazzani, M.J., Billsus, D.: Content-based recommendation systems. In: Brusilovsky, P., Kobsa, A., Nejdl, W. (eds.) The Adaptive Web. LNCS, vol. 4321, pp. 325–341. Springer, Heidelberg (2007). https://doi.org/10.1007/978-3-540-72079-9_10

20. Sandvig, J.J., Mobasher, B., Burke, R.D.: A survey of collaborative recommendation and the robustness of model-based algorithms. IEEE Data Eng. Bull. **31**(2), 3–13 (2008)
21. Stefanidis, K., Kondylakis, H., Troullinou, G.: On recommending evolution measures: a human-aware approach. In: 33rd IEEE International Conference on Data Engineering, ICDE 2017, San Diego, CA, USA, 19–22 April 2017, pp. 1579–1581 (2017)
22. Stefanidis, K., Ntoutsi, E., Kondylakis, H., Velegrakis, Y.: Social-based collaborative filtering. In: Alhajj, R., Rokne, J. (eds.) Encyclopedia of Social Network Analysis and Mining, pp. 1–9. Springer, New York (2017). https://doi.org/10.1007/978-1-4614-7163-9_110171-1
23. Stratigi, M., Kondylakis, H., Stefanidis, K.: Fairness in group recommendations in the health domain. In: ICDE (2017)
24. Stratigi, M., Kondylakis, H., Stefanidis, K.: FairGRecs: fair group recommendations by exploiting personal health information. In: Hartmann, S., Ma, H., Hameurlain, A., Pernul, G., Wagner, R.R. (eds.) DEXA 2018. LNCS, vol. 11030, pp. 147–155. Springer, Cham (2018). https://doi.org/10.1007/978-3-319-98812-2_11
25. Stratigi, M., Nummenmaa, J., Pitoura, E., Stefanidis, K.: Fair sequential group recommendations. In: Proceedings of the 35th ACM/SIGAPP Symposium on Applied Computing, SAC 2020 (2020)
26. Tamber: Steam video games. https://www.kaggle.com/tamber/steam-video-games/. Accessed 11 Apr 2019
27. Troullinou, G., Kondylakis, H., Stefanidis, K., Plexousakis, D.: Exploring RDFS KBs using summaries. In: Vrandečić, D., et al. (eds.) ISWC 2018. LNCS, vol. 11136, pp. 268–284. Springer, Cham (2018). https://doi.org/10.1007/978-3-030-00671-6_16
28. Yin, Z., Gupta, M., Weninger, T., Han, J.: LINKREC: a unified framework for link recommendation with user attributes and graph structure. In: WWW (2010)
29. Zhang, J., Peng, Q., Sun, S., Liu, C.: Collaborative filtering recommendation algorithm based on user preference derived from item domain features. Phys. A **396**, 66–76 (2014)

Using Twitter Streams for Opinion Mining: A Case Study on Airport Noise

Iheb Meddeb, Catherine Lavandier, and Dimitris Kotzinos[✉]

ETIS Lab, UMR 8051, CY Cergy Paris University, ENSEA, CNRS,
2 Avenue A. Chauvin, 95000 Pontoise, France
{Iheb.Meddeb,Catherine.Lavandier,Dimitrios.Kotzinos}@u-cergy.fr

Abstract. This paper proposes a classification model for opinion mining around airport noise based on techniques such as event detection and sentiment analysis applied on Twitter posts. Tweets are retrieved using the Twitter API either because of location or content. A dataset of preprocessed, with NLP techniques, tweets is manually annotated and then used to train an SVM (Support Vector Machine) classifier in order to extract the relevant ones from the obtained collections. The extracted tweets from the SVM classifier are fed to a lexicon-based classifier to filter out the false relevant and to increase precision. A lexicon-based sentiment classifier is then applied in order to separate positive, negative and neutral tweets. The sentiment classifier uses emoticons, polarity of words with subjective intensity, intensifiers, negation effect with dynamic scope, contrast effect and SWN to detect the sentiment of tweets in a hierarchical manner. The information present in the classified tweets is used for a statistical survey-like study.

Keywords: Twitter · Opinion mining · Natural language processing · Machine learning · Sentiment analysis · Text mining

1 Introduction

Microblogging has become a very popular communication intermediary these last years, such as Twitter [3], Tumblr [2], etc. Offering a social network service for people, they use it to share daily news and express their opinions or emotions towards several topics, in a completely free manner. In fact, Twitter has reached 336 million active users in the first quarter of 2018, according to Statista [1], and sharing around 500 million tweets per day. These numbers indicate the big amount of information shared and rapidly spread due to Twitter characteristics that enables 280 maximum characters in a post and introduces hashtags and usernames tagging. All of this has encouraged research in the field of data mining and natural language processing (NLP) to exploit microblogging services and especially Twitter. Different works have taken place in this context but aiming at different objectives. In our project, we aim to capture tweets shared by users who live in the area of an airport and discuss about noise problems generated by

© Springer Nature Switzerland AG 2020
G. Flouris et al. (Eds.): ISIP 2019, CCIS 1197, pp. 145–160, 2020.
https://doi.org/10.1007/978-3-030-44900-1_10

both air and road traffic and due (or not) to the presence of the airport and to understand their perception on the quality of life in the area. Heathrow airport is taken as an example to work with as it is one of the busiest airports and located in a highly and densely populated area.

The main goal of this project is to build a customizable platform that collects the stream of relevant tweets generated by users, store them and do the sentiment analysis. This wealth of expressed opinions though comes with a price: not all opinions, posts, discussions are relevant to a specific subject so we need first to be able to extract the relevant posts or discussions. This is not a trivial subject by itself, since the definition of a subject is not exact and the way people express themselves varies greatly. Moreover, the case of Twitter and other microblogging services is more complicated since their limit in the number of characters for each post forces people to express themselves in unique and sometimes difficult to decipher ways. So this led us to create ways to collect data automatically using information retrieval, data mining and machine learning techniques to extract the relevant posts. Additionally, we used sentiment analysis techniques in order to analyze the opinions expressed in tweets and extract the sentiment (positive, negative or neutral) involved. We hope to be able to offer an alternate method to the traditional surveying methods with an automatic and timely way. This faces several challenges such as dealing with trivial tweets, incomplete sentences, misspelling and abbreviation due to strictly short messages. Sentiment classification is a hard challenge that faces contextual meanings of messages such as irony and the use of emotional expressions. Our work can be used to survey opinions on different aspects of people's everyday lives but the Machine Learning (ML) algorithms we use, will need to be retrained in order to achieve reasonable results. So while this is not an out of the box approach, it is a complete effort to support online surveying on non-trivial subjects.

The rest of this paper is organized as follows. In Sect. 2, a study of the state of the art and related work is presented. Section 3 describes the proposed approach and the workflow for extracting sentiments about noise and quality of life from tweets. Experiments and results are shown in Sect. 4. Section 5 is the conclusion of this work and discussion of future perspectives.

2   State of the Art

2.1   Machine Learning Approaches for Sentiment Classification

Related works have mostly used emoticons [12], slangs and acronyms [11], words in text and their respective part-of-speech (POS), which is the grammatical description of word (e.g. noun, verb, adjective, etc.), intensifiers such as all caps and characters repetitions (e.g. happpyyy) [14], punctuation marks, n-grams and negation mark as features of tweets. The sentiment polarity of a tweet is, then, calculated using machine learning approaches or lexicon-based approaches.

According to [17], there are two classifier models, a 2-way and a 3-way sentiment classification. The 2-way model classifies texts into positive or negative and the 3-way model includes a neutral class with the previous ones. [12] showed that emoticons have a significant indication on the polarity of texts with a 2-way classification and emoticon-trained SVM (Support Vector Machine) [8] and Naive Bayes (NB) [10] classifiers were able to have more than 70% accuracy. However, this method has a poor performance with a 3-way classification. [17] tested the impact of n-grams on the classifier performance. They used NB for classification and showed that using bigrams leads to the best accuracy as it provides a good trade-off between a word meanings (unigram) and capturing sentiment expressions (trigrams). They also revealed that attaching negation words when using n-grams has a high accuracy even with a small training set. [14] used collections of hashtagged tweets and tweets with emoticons to see how useful features are. They took n-grams as baseline feature, and then tried combinations of it with a dictionary of subjective lexicon, POS features such as counting of verbs, nouns, adjectives and microblogging features (e.g. intensifiers, emoticons, slangs and abbreviations). They showed that applying all features together does not lead to the best performance but it depends on the type of features. Tree kernel is also a useful method to represent tweets [4] because polar (positive/negative) and non-polar (counts) features can be easily extracted. They also detect emoticons, negation and exclamation marks, stop and non-English words within the tree kernel. Their study showed that tree kernel combined with sentiment features (e.g. positive/negative words, count and prior polarity of POS, emoticons, etc.) outperforms the base line unigram. It is also important to mention that they took into account the subjective intensity of emoticons (e.g. extremely positive, positive, negative, etc.) but not those of words. Same as [14], they used combination of features to get the most effective ones. Their feature analysis showed that combining the prior polarity words with their POS gives the best performance, contrarily to [14]. This may be explained by the tagger errors and the use of POS in [4] (prior polarity of words by POS) and in [14] (count of POS).

[18] has used a context-based convolutional neural network (CNN) to apply sentiment classification on Twitter corpus with 5 main layers: tweets are represented by word embedding vectors to be passed, then, to the input layer. The convolution layer extracts lexical n-grams information and a max, min and average-pooling layer is used to know how important an n-gram is. They also used as sub network to extract contextualized words form tweets which were represented using tf-idf. A hidden layer is used to concatenate the values from the pooling layers of the main network and the sub network, which leads to the final output layer to get the polarity of tweets. They tested their model on tweets extracted from conversations, tweets sorted by author and tweets sorted by topic. Their study showed that their model gives the best performance on topic-based tweets.

2.2   Lexicon-Based Approaches for Sentiment Analysis

Besides machine learning approaches for sentiment classification, previous works have also used lexicon-based approaches that imply the use of dictionary of subjective words. For this purpose, many dictionaries from previous sentiment analysis already exist and research continues to take advantage of them because the creation of lexicon datasets is a time consuming task. Other than lexicon dictionaries, sentiment research works on microblogging messages have also used sets of positive and negative emoticons to detect sentiment classes, despite the fact that subjective words can be interpreted differently from one annotator to another. Moreover, even if the contents of the dictionaries (words) can be the same, their polarity might differ. To avoid these problems, [20] indicates the need of having more than one dataset to take into account multiple subjective perspectives of the word and to modify the existing dictionary, when necessary, to satisfy the topic sentiment characteristics or to create a domain specific dictionary using lexicon expansion techniques. [5] proposed a lexicon enhanced sentiment classifier on reviews to improve classification performances. In fact, they calculated the scores of positive and negative emoticons and words. The polarity score of a word is calculated using SentiWordNet classifier (SWNC) and a domain specific classifier (DSC) that takes into account the polarity of domain specific words both existing or unknown in SWNC. They also take into account negation (inverting the polarity score of the word next to the negation word) and modifiers, which are a sort of positive and negative grammatical intensifiers such as very, slightly, less, extremely, etc. They assign an intensity percentage to every modifier that represent its effect on the next word. The score of a sentence in a review is the summation of emoticons, modifiers, DSC and SWNC scores. Then a review is classified as positive, negative or neutral depending on the summation of sentences sentiment scores. Their study shows that DSC and modifiers have the best effect on improving performance and that DSC is used to give a correct classification of the misclassified neutral reviews due to the domain specific words that are nonexistent in SWNC so given a score of 0 (neutral).

2.3   Hybrid Classification Models

[13] also presented an hybrid sentiment classification framework on Twitter data. They used three different classifiers: emoticon classifier (EC), improved polarity classifier (IPC) and SWNC. Contrarily to [5], they detect the polarity of a tweet using a sequential method: After preprocessing tweets, they are passed to EC, which has positive and negative sets of emoticons. Depending on the emoticons in a tweet, EC classifies them into positive or negative. If tweet has a neutral score (i.e. does not have emoticons), it is passed to IPC which has sets of positive and negative words build from multiple existing lexicons datasets. Same to EC, the polarity of a tweet is calculated but this time, depending on words. If it is still neutral, the tweet is passed to SWNC. This algorithm has showed a good performance on classifying tweets especially on reducing the number of neutral tweets. However, they do not take into account the subjective intensity of words, negation nor modifiers.

3   Workflow for Extracting Sentiments from Tweets

Our proposed approach is presented as a workflow, which is divided into four main parts. First, queries are sent to Twitter Streaming API to collect tweets.

As the geographic area of our study is known (Heathrow airport). So we are collecting tweets using a location query to get messages within that area and also using a keywords query to get messages around Heathrow and aircraft noise. Then, messages are preprocessed using NLP techniques such as stop words removal, spelling correction, lemmatization, POS tagging, tokenization, etc. Afterwards, a machine learning algorithm, trained on an annotated dataset, is set up to filter out the irrelevant tweets and get the relevant ones. A domain knowledge classifier, which is lexicon-based, is also used to filter out irrelevant tweets. Relevant tweets are then preprocessed again because the first preprocessing task is only suited for relevance classification and does not satisfy sentiment classifier requirements The sentiment classifier uses sets of positive and negative emoticons, positive and negative lexicon with subjective intensity, and SWN to calculate the sentiment scores of tweets and to classify them into positive, negative or neutral. The use of these three classifiers is done in a hierarchical way by applying weights on their scores to have better performances.

3.1 Gathering Data: Twitter API

Twitter provides an API[1] to allow developers and researchers to access the publicly available user posts. They allow getting real time tweet streams with filtering by keywords, locations, languages, users, etc. the received tweet is represented as a JavaScript object notation (JSON) object that carries a lot of information about the tweet such as creation time, text, user description and location.

Retrieving Tweets with Location Query (TWLQ). Firstly, we define the area around Heathrow airport in which people will be talking about aircraft noise. We use the airport day, evening and night level (L_{den}) noise contours [6] to set the minimum surface of the area. We end up by defining a bounding box of 167 km wide, 73 km long and centered in Heathrow airport. The coordinates of the bounding box are introduced as a filter to Twitter API that is also configured to extract only English language tweets.

Retrieving Tweets with Keywords Query (TWKQ). The previous method gives only tweets having location, which are a small proportion of the overall accessible tweets (i.e. it misses a large number of relevant tweets that do not have a location). Moreover, it returns all tweets within that area so we get tweets talking about everything, which makes it impossible to take a sample with significant number of relevant ones for training. Therefore, we also use keywords queries to extract relevant tweets. We use "Heathrow", "LHR" and "noise" as keywords in a certain way to get tweets that have the words Heathrow and noise or LHR[2] and noise in the text.

---

[1] https://developer.Twitter.com/en/docs (Accessed on 08/17/2018).
[2] Airport code for London Heathrow.

3.2 Preprocessing Tweets (NLP)

Preprocessing tweets is an essential task for relevance classification and sentiment analysis. After retrieving tweets, URL links, numbers, emoticons and Twitter special words such as RT (denotes retweet) are removed. We keep usernames and hashtags as they can be informative features for relevance classification. Then the text is set to lowercase to ensure homogeneity of the following operations: Tokenization is applied to form a bag of words. Spelling errors within text are reduced by correcting intensified words (e.g. "happyyyy" becomes "happy"). Then, a POS tag is assigned to each word and the stop words are removed. Finally, lemmatization is applied to get a bag of root words that defines a tweet along with its usernames and hashtags. The preprocessed tweets will be used for relevance classification, which extracts relevant texts to be used for sentiment classification. However, this set of tasks is not very effective for sentiment analysis as they represent more the topic by the root words and so, loses the sentiment of sentences. Moreover, doing all the preprocessing in one step is not a desirable solution since the number of relevant tweets is much smaller than the number of irrelevant tweets. So the relevant tweets are preprocessed again, but differently; it starts with extracting emoticons and hashtags from text to be used later, followed by removing URL, usernames and punctuation marks. The symbol "#" is also removed from hashtags and we correct those who are composed by multiple words (e.g. hashtags "#NoisePollution" or "#noise_pollution" become "noise pollution") because words in hashtags can also be involved in the tweet's sentiment. However, the position of hashtags is not taken into account as we add all modified hashtags at the end of the tweet. The text is then set to lower case and tokenized. We use, as in the first step, the same spelling correction on each word but also detecting intensifiers such as character repetition and all caps. Words are then POS tagged and negation marks (e.g. not, 't and no) are detected. In that case, a negative mark is assigned to each of their following words. It is important to know where negation effect stops. In our case, the assignment gets back to normal when a sentence in a tweet ends. In microblogging messages, "," and "-" can also be used to end or start sentences besides normal ones such as points, exclamation and question marks. The negation scope also stops when conjunctions like "and", "or", wh-determiners (e.g. that, which), wh-pronouns (e.g. what, who), wh-adverbs (e.g. where, when) or contrast (e.g. but, however) words are found [9]. We also detect contrast in tweets as they have an effect on determining sentiments. The sets of emoticons, words with their POS and normal/negative effect and intensifiers are passed to the sentiment classifier.

3.3 Relevance Classification

After the first preprocessing part, tweets are set to be in the form of bag of root words and hashtags. We take unigrams, bigrams and hashtags as features and we used tf-idf technique to represent tweets. SVM algorithm is trained on an annotated sample of tweets, which are taken from the retrieved datasets TWLQ and TWKQ. The relevant classified tweets from SVM are introduced to a lexicon-based classifier. This classifier uses datasets of domain knowledge unigrams, bigrams and related hashtags and usernames, which were created from manually labeled relevant tweets, to calculate a domain knowledge score of each tweet. Then, the lexicon-based algorithm classifies a tweet as relevant when its relevance score is over a threshold ϵ. Else, the tweet is classified as irrelevant. ϵ is user or experimentally defined and is application specific. Figure 1 describes the flowchart of relevance classification. The threshold ϵ is set to be low to have a small impact on missing more tweets that are relevant but an important impact on reducing the number of false relevant. This method helps to filter out false relevant tweets and to have more classification precision as relevance results will affect the results of our sentiment analysis later.

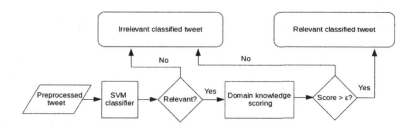

Fig. 1. Relevance classification flowchart

3.4   Sentiment Analysis of Relevant Tweets

After extracting relevant tweets from the stream, the appropriate preprocessing tasks are applied on relevant tweets. The proposed approach classifies them by their positive, negative or neutral sentiment using emoticons (Em), lexicon polarity (LP) of words and SWN. Let RT be the set of relevant tweets rt, W be the set of words w including the preprocessed hashtags and E be the set of emoticons e extracted from a tweet such as:

$$RT = \{rt_1, rt_2, ..., rt_j, ..., rt_n\} \tag{1}$$

$$W = \{w_1, w_2, ..., w_j, ..., w_m\} \tag{2}$$

$$E = \{e_1, e_2, ..., e_j, ..., e_t\} \tag{3}$$

Therefore, a relevant tweet rt is defined by:

$$rt = \{W, E\}, (rt \in RT) \tag{4}$$

Emoticon (Em) Score Calculation. Emoticons are extracted from tweets using regular expressions. We extract emoticons that are represented by punctuation marks or by Unicode. We created two datasets of positive emoticons PE and negative emoticons NE stored in files. The datasets have 64 emoticons divided into 38 positive and 26 negative. Sentiment scores of emoticons in a tweet rt_j are normalized and scaled between 1 and -1 such as:

$$score_{Em}(rt_j) = \frac{\sum_{i=1}^{t} emscore(e_i)}{t}, \ (e_i \in E) \wedge (E \in rt_j) \tag{5}$$

And:

$$emscore(e_i) = \begin{cases} 1 & \text{if } (e_i \in PE) \\ -1 & \text{if } (e_i \in NE) \\ 0 & \text{if } (e_i \notin PE) \wedge (e_i \notin NE) \end{cases} \tag{6}$$

Lexicon Polarity (LP) Score Calculation. LP score calculation is based on datasets of positive and negative words. Datasets are created from multiple existing lexicon collections to expand them and to avoid misinterpretation of sentiment of certain words. Lexicon lists from Liu [15], McDonald [16] and MPQA [19] are used to create the dataset. Duplicates and words that do not have the same polarity within all datasets are removed. We have also added, when missing, some of the domain knowledge subjective words from the work in [7] such as deafening, awake, unbearable, etc. And we removed the words "noise" and "noises" because they appear in most of the tweets so they would wrongly affect the sentiment polarity. Table 1 presents the statistics of positive and negative words from each resource and those that are used. Let PW denotes the set of positive words and NW the set of negative words. As subjective intensity of words is defined in [19], we used this intensity in scoring and we set the subjective intensity of additional words from the other dataset to unknown. So the dataset has 3 descriptions of subjective intensity of words: strong subjectivity, weak subjectivity and unknown subjectivity. We also set the domain knowledge subjective words to have strong subjectivity and changed the polarity of some related words to be suited for our topic. For example, we set "low" to have negative polarity as low flying planes cause more noise. Since intensifiers, negation and contrast words are detected. We use other additional sets for the scores. Let ACI be the set of all caps intensifier scores aci and CRI be the set of character repetition intensifier scores cri of each word such as:

$$ACI = \{aci_1, aci_2, ..., aci_j, ..., aci_m\} \tag{7}$$

$$CRI = \{cri_1, cri_2, ..., cri_j, ..., cri_m\} \tag{8}$$

Table 1. Statistics of lexicon datasets

Words	Datasets				
	Bing Liu	Bill McDonald	MPQA	Clashes and duplicates	Final dataset
Positive	2006	347	2719	6548	3251
Negative	4780	2306	4919		7278

If one of these intensifier is detected in a word, its following score will be 1.5 and 1 if it is not. For example the sets ACI and CRI of the tweet "plane noise is LOUD tonight! Respiiiiite #NOIIISE" will be $\{1, 1, 1, 1.5, 1, 1, 1.5\}$ and $\{1, 1, 1, 1, 1, 1.5, 1.5\}$ respectively. Let NEG and CON be the sets of negation and contrast words respectively. As the algorithm detects the negation, contrast and negation stop marks from the preprocessing part, normal, negative or inverse effect is assigned to each word using keywords. The normal sentiment score $swscore_{LP}$ of a word w_j in a tweet rt_j is:

$$swscore_{LP}(w_j) = \begin{cases} 1 \times weight \times aci_j \times cri_j & \text{if } (w_i \in PW) \\ (-1) \times weight \times aci_j \times cri_j & \text{if } (w_i \in NW) \\ 0 & \text{if } (ur_i \notin PW) \wedge (ur_i \notin NW) \end{cases}$$

$$(9)$$

where $weight$ is the subjective weight of the word. Its multiplication with aci_j and cri_j indicates the impact of the word on the tweet sentiment score and its polarity. When a word has a negative effect due to negation words, its score is multiplied by -1, allowing its opposite effect to be counted rather than its normal effect. So the sentiment score of the word, in this case, will be:

$$swscore_{LP}(w_j) = (-1) \times swscore_{LP}(w_j) \qquad (10)$$

This score is valid for all the words following the negation mark until a negation stop word is found or the sentence in a tweet ends. The LP score is calculated in an iterative manner, initializing it to zero and adding each time the score of the word such as:

$$LPscore = LPscore + swscore_{LP}(w_j), \; (w_j \in W) \wedge (W \in rt_j) \qquad (11)$$

When a word has an inverse effect, which means it is a contrast word, the following part of the sentence often has an opposite meaning of the first part and it also indicates the overall sentiment toward a subject. So, when a word such as "but" and "however" is found in a tweet, the polarity of the current score is inverted:

$$LPscore = (-1) \times LPscore \qquad (12)$$

This allows us to take into account the opposite meaning of sentence before the contrast word. After inverting the polarity, the algorithm continues to add scores of words normally. When another contrast word is found in the same tweet, the polarity will be inverted again. When all the polarity scores of words in a tweet are calculated and added to $LPscore$, it is normalized to ensure the sentiment intensity of a tweet:

$$score_{LP}(rt_j) = \frac{LPscore}{m}, \; (rt_j \in RT) \qquad (13)$$

SentiWordNet (SWN) Score Calculation. SWN dictionary is used for this purpose. In fact, each word in the dictionary have a positive, a negative and a neutral score, with a total score of 1. Its scores also depend on its POS tag and so, how it is employed in a text. Each word w_j in a tweet rt_j is introduced, with its POS tag to SWN to get also its synsets, which are words having the same meaning of w_j in a particular POS, to be counted in the word polarity scoring such as:

$$syscore_{SWN}(sy_i) = posscore_{SWN}(sy_i) - negscore_{SWN}(sy_i), (sy_i \in SY_{w_j}) \quad (14)$$

where SY_{w_j} is the set of synsets of the word w_j:

$$SY_{w_j} = \{sy_1, sy_2, ..., sy_j, ..., sy_v\} \quad (15)$$

And $posscore_{SWN}$ and $negscore_{SWN}$ are positive and negative scores of a word in SWN respectively. This enables us to take into account, same as LP scoring, the sentiment intensity of a word, however, the scores are not related to the topic. After calculating the score of each synset, we take their average to get the sentiment score of the word w_j such as:

$$swscore_{SWN}(w_j) = \frac{\sum_{i=1}^{v} syscore_{SWN}(sy_i)}{v}, (sy_i \in SY_{w_j}) \wedge (w_j \in W) \quad (16)$$

And the sentiment score of a tweet rt_j by SWN scoring method is:

$$score_{SWN}(rt_j) = \frac{\sum_{i=1}^{m} swscore_{SWN}(w_i)}{m}, (w_i \in W) \wedge (W \in rt_j) \wedge (rt_j \in RT) \quad (17)$$

Sentiment Score Calculation and Classification. The sentiment analysis approach uses the three scoring methods to determine sentiment polarities of tweets. They are used in a hierarchical way using weightings of scores and priority steps. Figure 2 shows the flowchart of sentiment classification algorithm. Firstly, emoticons and LP scoring algorithms are used to identify the sentiment of a tweet rt_j such as:

$$score_{Em+LP}(rt_j) = wei_i \times score_{Em}(rt_j) + wei_2 \times score_{LP}(rt_j) \quad (18)$$

where wei_1 and wei_2 are the weights assigned to Em and LP respectively. The classifier detect the sentiment of a tweet on the basis of thresholds. Let θ_1 and θ_2 be the respective positive and negative thresholds close to zero. If $score_{Em+LP}$ of a tweet rt_j is higher than θ_1, it is classified as positive and it is classified as negative if $score_{Em+LP}$ is lower than θ_2. Otherwise, if the score is between θ_1 and θ_2, the sentiment class of the tweet is not defined yet and it is fed to SWN scoring algorithm. Let $sclass_{Em+LP}(rt_j)$ be the sentiment class of the tweet on the basis of Em and LP scores such as:

$$sclass_{Em+LP}(rt_j) = \begin{cases} \text{positive} & \text{if } score_{Em+LP}(rt_j) > \theta_1 \\ \text{negative} & \text{if } score_{Em+LP}(rt_j) < \theta_2 \\ score_{SWN}(rt_j) & \text{if } score_{Em+LP}(rt_j) \in [\theta_2, \theta_1] \end{cases} \quad (19)$$

This reduces the number of tweets that are misclassified as neutral. Same as before, the sentiment classification of tweets on the basis of SWN score is done, using thresholds. Let τ_1 and τ_2 be thresholds close to zero. The sentiment class $sclass_{SWN}(rt_j)$ of a tweet rt_j on the basis of SWN is:

$$sclass_{SWN}(rt_j) = \begin{cases} positive & if\ score_{SWN}(rt_j) > \tau_1 \\ negative & if\ score_{SWN}(rt_j) < \tau_2 \\ neutral & if\ score_{SWN}(rt_j) \in [\tau_2, \tau_1] \end{cases} \tag{20}$$

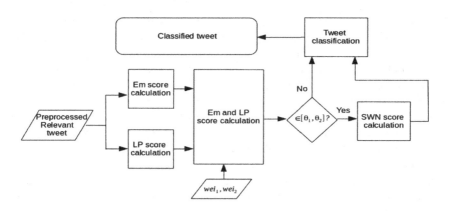

Fig. 2. Sentiment classification flowchart

4   Experimental Results

We have implemented all the workflow described in the previous section in the python programming environment. The confusion matrix, is used to analyze data and metrics such as *precision*, *recall* and $F - measure$ for each class and the *accuracy* are used to evaluate the classifiers performances. The confusion matrix is defined in Table 2. For example, *precision*, *recall* and $F - measure$ of the positive class are defined as follows:

$$precision_{pos} = \frac{T_{pos}}{T_{pos} + F_{pos\_neg} + F_{pos\_neu}} \tag{21}$$

$$recall_{pos} = \frac{T_{pos}}{T_{pos} + F_{neg\_pos} + F_{neu\_pos}} \tag{22}$$

$$F - measure_{pos} = 2 \times \frac{precision_{pos} \times recall_{pos}}{precision_{pos} + recall_{pos}} \tag{23}$$

And the accuracy is:

$$accuracy = \frac{T_{pos} + T_{neg} + T_{neu}}{All\_tweets} \tag{24}$$

Table 2. Confusion matrix

Confusion matrix		Predicted class		
		Positive	Negative	Neutral
Known class	Positive	T_{pos}	F_{neg_pos}	F_{neu_pos}
	Negative	F_{pos_neg}	T_{neg}	F_{neu_neg}
	Neutral	F_{pos_neu}	F_{neg_neu}	T_{neu}

Relevant tweets from the datasets TWLQ and TWKQ are taken to create a new dataset D_1. Tweets were labeled as positive, negative and neutral and were used for testing. Some details of D_1 are given in Table 3. As, generally, nobody likes being affected by aircraft noise and shows happy emotions towards airport noise, we defined a tweet as positive when the user shows a contrary opinion to negative tweets (e.g. "I live 10 min away from Heathrow. Noise is not disturbing, there is no air pollution.") and a tweet as neutral when it does not show a sentiment toward the topic or does not refer to airport noise (e.g. "Daytime aircraft noise was defined as that occurring between 0700 and 2300 h, and that occurring between 2300 and 0700 h was defined as night-time aircraft noise"). The numbers of positive and neutral tweets in the corpus, as showed in Table 3, are very small compared to the number of negative tweets (601 negative and 26 for each positive and neutral). Something expected, as most of the people have negative sentiments toward airport noise.

4.1   Sentiment Classifiers Comparison

We have studied the performance of the proposed classifier (PC) compared to emoticon classifier (EmC), LP classifier (LPC) and SWNC. We tuned thresholds to be suited for each classifier and fed tweets from D_1 to be classified. We set $wei_1 = 0.7$ and $wei_2 = 0.3$ so setting Em to have the priority over LP to classify tweets, when emoticons are present. We also set the weights of words that have strong subjectivity to 1, weights of words that have weak and unknown subjectivity to 0.75. The classifiers' performance is evaluated by calculating their respective confusion matrix and metrics. The results are given in Table 4 and show that our classifier outperforms the other classifiers. In fact, EmC has the worst results, having only 4.90% accuracy. Since this classifier only relies on emoticons to detect tweets sentiments, it classifies all the tweets that haven't emoticons as neutral. Table 3 shows that only 18 tweets in D_1 have emoticons and not all emoticons are recognized as some of them are neutral and others are not included in emoticon lists PE and NE. So the rest of tweets are automatically classified neutral which leads to have a big number of F_{neu} and consequently, a weak $F - measure$ and $accuracy$. However, it's still a good classifier when emoticons are present in a tweet because it captures negative tweets and shows a good $precision$. It misclassifies some of the negative tweets because of irony.

LPC has better results than EmC, with 62.17% $accuracy$ and a good $recall$ for negative and neutral classes (63.23% and 61.54% respectively) and

Table 3. Statistics of D_1

	Positive tweets	Negative tweets	Neutral tweets	Total	Tweets with emoticons
D_1	26	601	26	653	18

less $recall_{pos}$. Moreover, it has the best $F - measure$ results of positive and neutral tweets among all the classifiers. However, the difference between the numbers of negative tweets and the number of positive and neutral tweets has an effect on the precision of LPC. In fact, the number of the misclassified negative tweets (i.e. F_{pos_neg} and F_{neu_neg}) is higher than the total number of positive and neutral ones, leading to have low precision of positive and neutral classes with 12.50% and 8.89% respectively. The proportions of positive, negative and neutral tweets in the corpus depend on the topic, so it is part of the problem and needs to be taken into account. The 155 misclassified negative tweets as neutral F_{neu_neg} are, principally, due to the missing sentiment words in the lexicon lists PW and NE and so, most of these tweets have a score of 0 and are within the thresholds interval, consequently, they are classified as neutral. On the other hand, the misclassified negative tweets as positive are due to multiple reasons but mainly, the use of contrast in score calculation, which it does not take into account the sentence level in a tweet. SWNC is better than LPC, with 7.20% more $accuracy$. However, $F - measure$ of positive and negative classes are lower than those of LPC resulting to low values of $precision$ and $recall$ for each one of them. This is due to the decrease of T_{pos} and T_{neu} tweets. Additionally, F_{neu_neg} has decreased and T_{neg} has increased, compared to LPC, which leads to the increase of $recall_{neg}$. PC has the best performance with 77.79% $accuracy$.

Table 4. Experiments and results of EmC, LPC, SWNC and PC on D_1

D_1			Confusion matrix			Metrics			
			Positive	Negative	Neutral	precision	recall	$F - measure$	accuracy
EmC	Thresholds	Positive	0	0	26	0%	0%	-	4.90%
	$\theta1 = \theta2 = 0$	Negative	3	6	592	100%	1%	1.98%	
		Neutral	0	0	26	4.04%	100%	7.76%	
LPC	Thresholds	Positive	10	7	9	12.50%	38.46%	18.86%	62.17%
	$\theta1 = 0.027$	Negative	66	380	155	96.69%	63.23%	76.46%	
	$\theta2 = -0.001$	Neutral	4	6	16	8.89%	61.54%	15.53%	
SWNC	Thresholds	Positive	4	15	7	4.44%	15.38%	6.90%	69.37%
	$\tau_1 = 0.015$	Negative	79	443	79	95.05%	73.71%	82.64%	
	$\tau_2 = 0.005$	Neutral	7	13	6	6.52%	23.07%	10.17%	
PC	Thresholds	Positive	12	10	4	12.12%	46.15%	19.20%	77.79%
	$\theta_1 = 0.01,$	Negative	80	491	30	95.34%	81.70%	87.99%	
	$\theta_2 = -0.001$								
	$\tau_1 = 0.015,$	Neutral	7	14	5	12.82%	19.23%	15.38%	
	$\tau_2 = 0$								

It also has a good *precision, recall* and $F - measure$ compared to the other classifiers. The architecture of PC enabled us to decrease, significantly, the number of tweets classified neutral and so the number of F_{neu_neg}. It firstly, uses LP and Em scores to have a good *precision* results on positive and negative classes. Secondly, F_{neu} is decreased by classifying all the unclassified tweets with SWN score algorithm which also leads to increase T_{neg} and T_{pos}. This method, however, increases F_{pos_neg} and decreases T_{neu} with the worst $recall_{neu}$ of 19.23% but it still keeps a good $F - meausre_{pos}$ and $F - measure_{neu}$ compared to the other classifiers with 19.20% and 15.38% respectively.

5   Conclusions

This paper presents the workflow for a solution for detecting tweets relevant to a specific subject and extract their sentiments. Noise around Heathrow airport is taken as an example to work with. Tweets are retrieved using Twitter API with two methods: the first with location filter (area around Heathrow airport) and the second with keywords filter ("Heathrow", "LHR" and "noise"). Tweets are then preprocessed using a combination of NLP techniques which is suited for relevance classification. Relevant tweets towards airport noise are then extracted from the stream using SVM and a lexicon-based classifier. Relevant tweets are then preprocessed again using other combination of NLP techniques to be suited, this time, for sentiment classification. The sentiment classifier uses emoticons, lexicon polarities with subjective intensity of words, negation effect with dynamic scope, intensified words and also contrast words and SentiWordNet scores in a hierarchical way to detect the sentiments of tweets and classify them into positive, negative or neutral sentiments.

Experimental results showed that the proposed classifier outperforms the emoticon classifier, the subjective lexicon-based classifier and SWN classifier. Moreover, it still has a margin for improvement as it captures significant number of false positive tweets. As perspectives, we suggest to improve the sentiment classifier by expanding the subjective lexicon. The spelling correction needs also to be improved by replacing slang words, correcting different types of misspelling. The polarity inverting feature due to contrast can also be improved by limiting the effect at the sentence level or only count the polarity of the sentence following the contrast word to avoid misclassifications. Sentiment classes can also be divided into normal, strong or weak sentiments. Grammatical intensifiers (e.g. very, more, less, extremely, quite) can also be taken into account in further works. Finally, we plan to apply the same methodology and validate the method followed on a number of different topics, so as to demonstrate its wider applicability to the problem of exploiting social media data in order to extract people's sentiment for a particular topic of interest.

Acknowledgements. This work has been partially supported by the ANIMA project, which has received funding from the European Union's Horizon 2020 research and innovation programme under grant agreement No 769627. Website: https://anima-project.eu/.

References

1. Statista. https://www.statista.com/statistics/282087/number-of-monthly-active-Twitter-users/. Accessed 13 Aug 2018
2. Tumblr. https://www.tumblr.com/. Accessed 13 Aug 2018
3. Twitter. https://Twitter.com/. Accessed 13 Aug 2018
4. Agarwal, A., Xie, B., Vovsha, I., Rambow, O., Passonneau, R.: Sentiment analysis of twitter data. In: Proceedings of the Workshop on Languages in Social Media, LSM 2011, pp. 30–38. Association for Computational Linguistics, Stroudsburg (2011). http://dl.acm.org/citation.cfm?id=2021109.2021114
5. Asghar, M.Z., Khan, A., Ahmad, S., Qasim, M., Khan, I.A.: Lexicon-enhanced sentiment analysis framework using rule-based classification scheme. PLoS ONE **12**(2), 1–22 (2017). https://doi.org/10.1371/journal.pone.0171649
6. Civil Aviation Authority: Heathrow airport 2016 summer noise contours and noise action plan. Technical report (2017)
7. Barbot, B., Lavandier, C., Cheminée, P.: Linguistic analysis of field surveys carried out around two French airports. Technical report (2007)
8. Cortes, C., Vapnik, V.: Support-vector networks. Mach. Learn. **20**(3), 273–297 (1995)
9. Farooq, U., Mansoor, H., Nongaillard, A., Ouzrout, Y., Qadir, M.A.: Negation handling in sentiment analysis at sentence level. JCP **12**(5), 470–478 (2017)
10. Harry, Z.: The optimality of Naive Bayes. In: Proceedings of Florida Artificial Intelligence Research Society Conference (FLAIRS), pp. 562–567. AAAI Press (2004)
11. Hutto, C.J., Gilbert, E.: VADER: a parsimonious rule-based model for sentiment analysis of social media text. In: Proceedings of the Eighth International AAAI Conference on Weblogs and Social Media. The AAAI Press (2014)
12. Jonathon, R.: Using emoticons to reduce dependency in machine learning techniques for sentiment classification. In: ACL the Association for Computer Linguistics, pp. 43–48 (2005)
13. Khan, F.H., Bashir, S., Qamar, U.: TOM: twitter opinion mining framework using hybrid classification scheme. Decis. Support Syst. **57**, 245–257 (2014)
14. Kouloumpis, E., Wilson, T., Moore, J.D.: Twitter sentiment analysis: the good the bad and the omg! In: Proceedings of the Fifth International AAAI Conference on Weblogs and Social Media. The AAAI Press, Barcelona (2011)
15. Liu, B., Hu, M., Cheng, J.: Opinion observer: analyzing and comparing opinions on the web. In: Proceedings of the 14th International World Wide Web conference (WWW-2005). ACM, Chiba (2005)
16. Loughran, T., McDonald, B.: When is a liability not a liability? Textual analysis, dictionaries, and 10-ks. J. Finan. **66**(1), 35–65 (2011). https://EconPapers.repec.org/RePEc:bla:jfinan:v:66:y:2011:i:1:p:35-65
17. Pak, A., Paroubek, P.: Twitter as a corpus for sentiment analysis and opinion mining. In: Proceedings of the seventh International Conference on Language Resources and Evaluation (LREC 2010), Valetta, Malta (2010). http://www.lrec-conf.org/proceedings/lrec2010/pdf/385_Paper.pdf

18. Ren, Y., Zhang, Y., Zhang, M., Ji, D.: Context-sensitive twitter sentiment classification using neural network. In: Proceedings of the Thirtieth AAAI Conference on Artificial Intelligence. AAAI Press, Phoenix (2016)
19. Theresa, W., Janyce, W., Paul, H.: Recognizing contextual polarity in phrase-level sentiment analysis. In: Proceedings of HLT-EMNLP-2005, pp. 347–354. The Association for Computational Linguistics, Vancouver (2005)
20. Yadollahi, A., Shahraki, A.G., Zaïane, O.R.: Current state of text sentiment analysis from opinion to emotion mining. ACM Comput. Surv. 50(2), 25:1–25:33 (2017)

A Platform Development for Multilingual Law Collection and Comparative-Law Support Services: ASEAN Laws as a Case Study

Vee Satayamas[1(✉)], Asanee Kawtrakul[1], and Takahiro Yamakoshi[2]

[1] Kasetsart University, Bangkok, Thailand
vee.sa@ku.th, ak@ku.ac.th
[2] Nagoya University, Nagoya, Aichi, Japan
yamakoshi@kl.itc.nagoya-u.ac.jp

Abstract. Lawmakers in the ASEAN countries need to investigate statutes of neighbor countries to draft consistent, uniform, and reasonable statutes. Moreover, the non-lawyers, who would like to invest or work oversea, should understand the statutes of the countries under consideration and compare the regulation requirements before making decision which country is good for investment or for working. This work proposes a platform for collecting and comparing laws. It consists of three modules: the first one is a Web crawling for gathering the statutes from ASEAN countries' law archives, the second module is Document preprocessing for extracting the regulations from each statute of each country and aligning them across the text, and the last module is a service with a tool for highlighting the relevant parts of text. This paper proposes to use existing text processing tools, such as, word/word-group segmentation and document section parsing, to use Wikidata's ontological concept for annotating those entities, and then align them across the text. However, there are two problems of concept selection, i.e. concept ambiguity and concept granularity. A near-threshold of maximum distance to the least common ancestor is computed for selecting a proper concept for entity alignment. This work did an experiment on Malaysia and Thailand's labor law to compare the minimum wages. By testing with a several of thresholds, the threshold value two gives the most proper concept where the precision and recall of related entities alignment are 48% and 67%, respectively.

Keywords: Multilingual legal documents collection · Automatic translation · Concept annotation · Platform for law comparison · Ontology-based entity alignment

1 Introduction

Lawmakers in the ASEAN countries need to investigate statutes of neighbor countries and treaties among the countries to draft consistent, uniform, and

© Springer Nature Switzerland AG 2020
G. Flouris et al. (Eds.): ISIP 2019, CCIS 1197, pp. 161–174, 2020.
https://doi.org/10.1007/978-3-030-44900-1_11

reasonable statutes. To do comparative law, the lawmaker should explore and compare the law of one country to that of another for a better understanding of the nature of the law in different countries. The key act in comparison is looking at one mass of legal data in relation to another and then assessing how the two portions of legal data are similar and how they are different [18]. In addition, end-users, who are not lawyers such as investors and migrant workers, need to understand for complying the statues of the countries under consideration and then to compare the regulation requirements before making decision which country is good for investment or for working. This paper, then, proposes a framework and a developed platform for statutes collecting from legal websites of each country in ASEAN and extracting the portions that related to regulations/rules/conditions from each statute and aligning them across the countries for making comparison.

To accomplish the law comparison task, there are four main problems, i.e., law collection problem, not (yet) machine-readable text problem, related entities alignment problem, and language barrier problem.

– Law collection problem: At the current stage, statutes are gathering through the legal websites by using web crawlers. Since different websites are implemented with different standard formats, so it causes difficulty in the crawling process. For example, Thai statutes data are represented with tree components for browsing. In contrast, Malaysian statutes data are represented with list sorted by id [2,4,5,8,9,11,14,15]. Furthermore, some websites, it requires mimicking humans interacting with the website for browsing the data, especially the website that constructed by using Java script.
– Not (yet) machine-readable text problem: Since almost all websites provide statutes data in PDF format, so most of law documents cannot be further processed for extracting the text portions and aligning those related content for making comparison.
– Entity alignment problem: In order to compare statutes of one country to that of another for consideration, the entities in the text portion should be annotated with concepts for supporting the alignment of related clauses/rules/conditions in the statute. This work utilizes the ontology from Wikidata as a concept for annotating the entities. However, there are still two sub problems of proper concept selection. For example, Phuket, a name entity of a province in Thailand, has three possible concepts, hereafter called concept ambiguity, i.e., a province, a city, an island, a film, visual art, or creative work. Moreover, Phuket can be labeled whether with more specific concept ancestor or more generic concept ancestor, hereafter called concept granularity, i.e., a province, administrative territorial entity, a human-geographic territorial entity, an artificial entity, or an entity which is the root of concepts, respectively.

- Language barrier problem: Many statutes of Indonesia [5], Laos [14], Vietnam [12], and Thailand [4] are not translated into English. Thus, they cannot be compared directly. Even, there is a law database that was designed for keeping statutes of all ASEAN countries [1], the number of available statutes in English is very few.

In order to solve the problems mentioned above, this work proposes a framework and developed a platform for providing a service for both lawmakers and non-lawyers who would like to review/study/understand the difference of rules in the statutes. It consists of three modules, i.e.,

- The first module is a Web crawling for gathering the statues from ASEAN countries' law archives. The virtual human-interaction is also implemented to download statutes.
- The second module is Document preprocessing which consists of 7 submodules, starting with converting not (yet) machine-readable content in PDF format by using Google's Optical Character Recognition API [10] and converting HTML format to be plain text by using Nokogiri [7]. The rest of submodules also utilizes the existing tools for parsing, extracting the regulations/rules/conditions from the parsed text and then aligning the pairs of related entities by using ontology from Wikidata for further translation processing and then providing the service in making comparison. The most challenges in document processing is that: many ASEAN languages still do not have an accurate part-of-speech tagger or an accurate syntactic parser for words or phrases segmentation and text translation. In addition, there are several candidate concepts for annotating the entities before making comparisons.
- The last module is to provide a service with a tool for highlighting the parts of text that the end users would like to pay attention to.

The remainder of this paper is organized as follows. Section 2 describes a platform for multilingual law collection and comparative-law support services. Section 3 describes the implementation and results. The conclusion and future direction is given in the last section.

2   A Platform Design for Multilingual Law Collection and Comparative-Law Support Services

The designed platform consists of three modules, as shown in Fig. 1: (1) web crawling, (2) document preprocessing, and (3) comparative law support services providing, which are described in Sects. 2.1, 2.2, and 2.3, respectively.

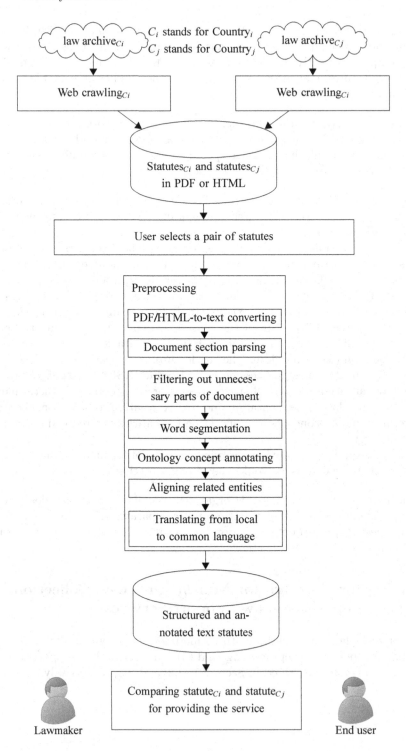

Fig. 1. A platform for multilingual law collection and comparative-law support services

2.1   Web Crawler for Harvesting Statutes

This paper reported the survey result of ASEAN countries' legislative information. Table 1 shows the list of surveyed websites of 10 countries, namely Brunei, Cambodia, Indonesia, Myanmar, Laos, Malaysia, the Philippines, Singapore, Thailand, and Vietnam. There are no online source of Cambodia and Myanmar. For the rest of websites, including Thailand and Malaysia, statutes can be downloaded manually.

Table 1. Online law archives survey

Country	Url	Status
Brunei	http://www.agc.gov.bn	Downloadable
Cambodia	Unknown	N/A
Indonesia	https://www.greengazette.id	Downloadable
Myanmar	Unknown	N/A
Laos	http://www.na.gov.la/	Available (Under construction)
Malaysia	http://www.agc.gov.my	Downloadable
The Philippines	https://www.officialgazette.gov.ph	Downloadable
Singapore	https://sso.agc.gov.sg/	Downloadable
Vietnam	http://vietnamlawmagazine.vn	Available (Subscription is needed)
Thailand	http://www.mratchakitcha.soc.go.th	Downloadable
	http://www.krisdika.go.th/	Downloadable

In order to collect the statues, automatically, from those accessible websites mentioned in Table 1, a web crawler has been developed. However, the task of statutes collection is not easy, since different websites are implemented with different user-interface design, i.e., JavaScript-based tree user interface, JavaScript-based list user interface, and HTML-based list user interface. Accordingly, the specific web crawler is needed for mimicking human interaction to those user interfaces through a script that controls Firefox [22] via SlimerJS [17]. The developed web crawler also run periodically for obtaining new and updated statutes.

2.2   Document Preprocessing

To use the platform, an end-user has to select a pair of statutes/announcements as inputs. Then, the system will execute the following seven steps: (1) converting PDF/HTML to text, (2) parsing the whole text to subsections, (3) filtering out unnecessary parts such as the introduction part, (4) word segmenting for concept labeling, (5) annotating by using ontological concepts from Wikidata, (6) aligning related entities across text through the annotated concepts, and (7) translating local languages to ASEAN common language, i.e., English.

Algorithm 1. Statute preprocessing

```
# DATABASE SERVICE: Wikidata
# INPUT: A pair of statutes (in PDF or HTML)
# OUTPUT: A pair of translated statutes with aligned entities
# STEP 1 Converting PDF/HTML
statutes_step1 = convert_html_pdf(a pair of statutes)
# STEP 2 Parsing document sections
statutes_step2 = parse_doc_sections(statutes_step1)
# STEP 3 Filtering parts of a statute
statutes_step3 = filter_document_parts(statutes_step2)
# STEP 4 Word segmentation
statutes_step4 = segment_words(statutes_step3)
# STEP 5 Ontology concept annotating
statutes_step5 = ontology_concept_annotate(statutes_step4, wikidata)
# STEP 6 Aligning related entities based on ontology concepts
statutes_step6 = align_entities(statutes_step5)
# STEP 7 Translating local languages to ASEAN common language
statutes_step7 = translate(statutes_step6)
```

Algorithm 1 mentioned above runs the following steps (1 to 4) by using the existing tools:

- Use Google's Optical Character Recognition API [10] to convert statutes in PDF format to be plain text format,
- Use Nokogiri [7] to convert statutes in HTML format to be plain text format,
- Use clue words, for parsing the whole text to subsections, for examples, "หมวดที่" (in Thai)/"Section" (in English), "มาตรา" (in Thai)/Aricle/Section (in English), and "สั่ง ณ" (in Thai) /"Orderd at" (in English),
- Use spaCy [3] for English word segmentation and English lemmatization,
- Use Chamkho [13] for Thai word segmentation,

The challenges of this work is how to select the proper concepts from Wikidata for concept labeling to those segmented words or word groups (hereafter called "entities"), which are the output from step 4. Two sub-tasks are needed for annotating the concepts to the entities, i.e., generating candidate entities, and querying ontology concepts.

In order to use "minimum wage" as a concept, while "Minimum Wages" is used in an announcement/statute, a candidate set of entities should be generated as Minimum, Wages, minimum, wage, Minimum Wages, and minimum wage.

Algorithm 2. Generating cadidate entities

```
INPUT: words, language
OUTPUT: entities
entities = []
for w in words:
  entities.append(w)
if language == :ENGLISH
  for word-bi-gram in generate-bigrams(words):
    entities.append(word-bi-gram)
  lemmas = [find-lemma(w, language) for w in words]
  for lemma-bi-gram in generate-bigrams(lemmas):
    entities.append(lemma-bi-gram)
return entities
```

In order to obtain Wikidata concepts for annotating every entity in the announcement, the system will generate SPARQL [19] statements by using Algorithm 3 for querying the related concepts from the library, called Mundaneum [6].

Algorithm 3. Wikidata query

```
(template [:select ?e
           :where [:union [[[?e skos:altLabel ~w@~lang]]
                          [[?e rdfs:label ~w@~lang]]
           :limit 10])))
```

The generated SPARQL statements for querying the concept of "Minimum Wages" are shown as the follows:

```
SELECT ?e WHERE { { ?e rdfs:label "Minimum Wages"@en. }
  UNION { ?e skos:altLabel "Minimum Wage"@en } }
SELECT ?e WHERE  { { ?e rdfs:label "minimum wage"@en. }
  UNION { ?e skos:altLabel "minimum wage"@en } }
```

By submitting the above queries to the Wikidata SPARQL service, wd:P6794 and wd:Q186228 will be returned. Where wd:P6794 is a property that can be used for retrieving a pair of values such as Germany and 9.35 EUR, and wd:Q186228 is an item that can be used for retrieving its properties and related values such as "subclass of", and "minimum wage in Quebec", as shown in Fig. 2.

Fig. 2. "Minimum wage" property and "minimum wage" item in their context

In step 6, both wd:P6794 and wd:Q186228 will be used as annotated concepts for "Minimum Wages" written in the announcement. Table 2 shows the other examples of querying concepts from Wikidata for two entities, i.e., ระยอง (Rayong: a name of a province in Thailand), and "Sarawak" (a name of a state in Malaysia).

Table 2. Wikidata concepts for Rayong and Sarawak

Entity	Concepts
ระยอง (Rayong)	Q335221 Rayong (a province)
Sarawak	Q170462 Sarawak (a state of Malaysia)
	Q1658411 Sarawak (a kingdom on northern Borneo)

In order to align the related entities across the selected statutes, Algorithm 4 was applied. Figure 3 shows the partial results of running the function *find_ancestors* to retrieve ancestors of all concepts in Table 2. For example, the ancestors of Q1658411 (Sarawak: a kingdom) are Q417175 (kingdom), Q1250464 (realm), Q7275 (state), Q43229 (organization), Q24229398 (agent), and Q35120 (entity).

However, to align related entities across statutes by using ancestor concepts still has two problems, namely, concept ambiguity and concept granularity. In case of annotating "Sarawak", there are two possible concepts. One is Q170462 (Sarawak: a state). Another one is Q1658411 (Sarawak: a kingdom). But the proper one is Q170462. Regarding to concept granularity, if too general concepts, e.g., Q35120 (entity), were used, all entities would be determined to be related each other. If too specific concepts, e.g., Q335221 (Rayong), Q170462 (Sarawak: a state), and Q1658411 (Sarawak: a kingdom) would be used instead, no entity was related. To solve both concept ambiguity and concept granularity, the least common ancestor of each annotated concept pair should be retrieved by using the function *least_common_ancestor*. Two entities will be aligned if the maximum distance of least common ancestor between their annotated concepts is not more than the preferred threshold. As written in Algorithm 4, if *distance(lca, concept0)* <= *THRESHOLD*, and *distance(lca, concept1)* <= *THRESHOLD* is true, then the entity pairs will be selected.

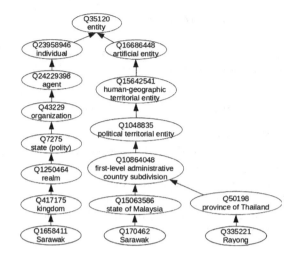

Fig. 3. Partial ancestors of Q335221 (Rayong), Q170462 (Sarawak: a state), and Q1658411 (Sarawak: a kingdom)

Algorithm 4. Aligning entities

```
INPUT: statute0, statute1, wikidata_ancestors_db
OUTPUT: related_entities
for entity0, concept0 in statute0:
    for entity1, concept1 in statute1:
        ancestors0 = find_ancestors(concept0, ancestor_db)
        ancestors1 = find_ancestors(concept1, ancestor_db)
        lca = least_common_ancestor(concept0, concept1, ancestor0, ancestor1)
        if distance(lca, concept0) <= THRESHOLD and \
                    distance(lca, concept1) <= THRESHOLD:
            related_entities.append([entity0, entity1])
```

In step 7, main sections of statutes would be translated from local languages to ASEAN common language, i.e., English, by using Google Translate, and Moses SMT [21]. We trained Moses SMT by using 5,726 Thai-English pairs from Thai law website [4], 2,881,915 Thai-English text unit pairs from Open Subtitle [23], and 77 Thai-English text unit pairs from GNOME L10N corpus [23].

2.3 Comparative Law Support Service

The output of translated statutes with related entities alignment will be highlighted by the system. However, there will be too many highlighted entity pairs. Therefore, we provide a tool for pruning unnecessary concepts by the end-users. More details explain in Sect. 3.3.

3   Implementation and Results

In order to prove the proposed framework and evaluate the platform, Malaysia and Thailand's labor law and its announcement are used as a case study.

3.1   Web Crawling Results

In order to gather the statutes/announcements from the Office of the Council of the State (Krisdika) website [4], the web crawler is developed to mimick a click for retrieving statutes through a tree-liked user interface. Consequently, all statutes and announcements listed on the website can be collected, i.e. 7,402 Thai statutes/announcements in HTML format, English translation of 129 Thai statutes/announcements in HTML format, and 423 Thai statutes/announcements in PDF format.

3.2   The Result of Document Preprocessing

The result of the document preprocessing on Malaysia and Thai minimum wage announcements preprocessing results are as follows. Google Cloud OCR can convert PDF into plain text without errors. Parsing text to document sections can decompose the Thai announcement to the introduction section, main section, and the ending section. Introduction section and ending sections are correctly filtered out. Thai and English word segmentations have no error. The system uses segmented words, word's lemma, and bigrams of words and lemma, for generating candidate entities. Ontological concepts are used for annotating every related entity such as *Sarawak* annotated with Q170462 (Sarawak: a Malaysian state), and *ระยอง (Rayong)* annotated with Q335221 (Rayong: a province in Thailand). In order to align related entities with the proper concepts, we conducted an experiment using different threshold's values, i.e., 1, 2, and 3. By testing with a several of threshold values, as shown in Table 3, using the maximum distance to the least common ancestor value 2, the alignment between related entities yields a balanced result, i.e., 48% precision and 67% recall.

Table 3. Comparing entities alignment by varying maximum acceptable distance from the least common ancestor

Threshold	Precision	Recall
1	60%	33%
2	48%	67%
3	27%	72%

Table 4. The results of correctly extracted pairs of related entities using threshold value 2

Entity from Malaysian announcement	Entity from Thai announcement	Least common ancestor
RM	เงิน (Translated word: money)	Q1368 (money)
Sabah	ชลบุรี (Chonburi, a province name)	Q10864048 (1st-level administrative country subdivision)
Sabah	ภูเก็ต (Phuket, a province name)	Q10864048 (1st-level administrative country subdivision)
...	...	...

Table 4 shows the examples of correctly extracted pair of entities when the threshold value 2 is used. Table 5 shows the cause of low precision even using threshold value 2. For example, Minimum in the wage/salary context should be annotated with *lower bound* concept, but Minimum is mistakenly annotated with "painting" concept. While ภูเก็ต (Phuket) in this wage context should be annotated with *province* concept. However, ภูเก็ต (Phuket) is mistakenly annotated as a film. Since "film" and "painting" concepts have a common ancestor as "visual art", consequently, Minimum and ภูเก็ต (Phuket) are inappropriately aligned. In order to solve concept ambiguity problem, those concepts, which are not related to the topic, such as a film, should be removed.

Table 6 shows the examples of entity pairs that should be aligned, but they were not because the maximum distance to the least common ancestor is 3 instead of 2. For example, *Peninsular Malaysia* and ชลบุรี (Chonburi) are not aligned, even the common concept is *administrative territorial entity*. Those unaligned concepts caused the low recall when using threshold value 2, i.e, 67% shown in Table 3. In the future, multi valued Threshold might be applied for some groups of entities.

To show the comparison in ASEAN common language, Thai statutes are translated into English by using Google Translate and Moses SMT. The Moses SMT was trained by aligned law documents from two open corpora, i.e., GNOME parallel corpus [23], and Open Subtitle parallel corpus [23]. Google Translate can translate 16 words of 16 words correctly, while Moses SMT failed to translate 11 words of 16 words. Therefore, Google Translate is preferred to Moses SMT for integrating with our system.

Table 5. The examples of incorrectly aligned pairs of entities due to concept ambiguity

Entity from Malaysian announcement	Entity from Thai announcement	Least common ancestor
800	วัน (Transliteration of *one*, or translated word: *day*)	Q21199 (natural number)
900	วัน (Transliteration of *one*, or translated word: *day*)	Q21199 (natural number)
Federal	ภูเก็ต (Phuket, a province name)	Q11424 (film)
Minimum	บาท (Translated word: THB)	Q17537576 (creative work)
Minimum	ภูเก็ต (Phuket, a province name)	Q4502142 (visual art)
RM	จังหวัด (Translated word: Province)	Q24017414 (first-order metaclass)
RM	ภูเก็ต (Phuket, a province name) Phuket)	Q2431196 (audiovisual work)
and	วัน (Transliteration of *one*, or translated word: *day*)	Q82042 (word class)
area	วัน (Transliteration of *one*, or translated word: *day*)	Q151885 (concept)
area	เป็น (Translated word: is)	Q18616576 (Wikidata property)

3.3 Comparative-Law Support Services

To provide a service for the end users, both lawmakers and non-lawyer, the system will highlight the related entities with different colors. However, there are too many highlighted entity pairs that are aligned. Therefore, the developed platform provides a tool for the end users to prune the uninteresting pairs.

Table 6. The examples of unaligned pairs due to maximum distance of least common ancestor is 3

Entity from Malaysian announcement	Entity from Thai announcement	Least common ancestor
Peninsular Malaysia	จังหวัด (Translated word: Province)	Q56061 (administrative territorial entity)
Peninsular Malaysia	ชลบุรี (Name entity of provice: Chonburi)	Q56061 (administrative territorial entity)
Peninsular Malaysia	ภูเก็ต (Name entity of provice: Phuket)	Q56061 (administrative territorial entity)
Peninsular Malaysia	ระยอง (Name entity of provice: Rayong)	Q56061 (administrative territorial entity)

Figure 4 shows the highlights when a user selects only three concepts that they would like to pay attention for discussing or for considering the parts that they concern.

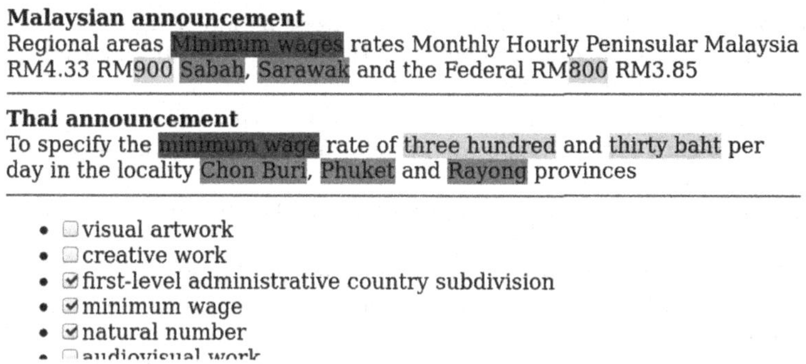

Malaysian announcement
Regional areas Minimum wages rates Monthly Hourly Peninsular Malaysia RM4.33 RM900 Sabah, Sarawak and the Federal RM800 RM3.85

Thai announcement
To specify the minimum wage rate of three hundred and thirty baht per day in the locality Chon Buri, Phuket and Rayong provinces

- ☐ visual artwork
- ☐ creative work
- ☑ first-level administrative country subdivision
- ☑ minimum wage
- ☑ natural number
- ☐ audiovisual work

Fig. 4. Law comparison service with a tool for highlighting the interesting pairs

4   Conclusion

The most challenge of this work is how to align the regulations across the statutes of ASEAN countries which almost are written in local language while there are poor resources for text processing, such as good text parsers and embedded text translation. Therefore, we study the feasibility of using the simplified tools for text processing, and annotating ontological concepts of Wikidata. To conduct the experiment, Malaysia and Thailand's labor law and its announcements are selected. With the developed platform, the web crawler that could mimick human interaction can collect all statutes available on the Office of the Council of the State (Krisadika). Document preprocessing can work effectively in extracting the regulations/rules/conditions from statutes among the interesting countries and in aligning them across the text for making comparison. In order to solve the two sub-problems, i.e., concept granularity and concept ambiguity, during aligning related entities, a near-threshold of maximum distance to the least common ancestor is computed for selecting a proper concept for entity alignment. For this work, we use the threshold value 2, which gave the precision and recall of related entity alignment, 48%, and 67%, respectively.

In the future, we plan to enhance the performance of web crawling by driving and promoting to use a linked data format, such as JSON-LD, RDF as a standard, for their law archive websites. To improve alignment, we will further study to find another technique such as word sense disambiguation using word embeddings [16,20].

References

1. ASEAN Legal Database (2019). http://asean-law.senate.go.th. Accessed 29 Mar 2019
2. Attorney General's Chambers (2019). http://www.agc.gov.bn/AGCSitePages/INDEXTOTHELAWSOFBRUNEI.aspx. Accessed 29 Mar 2019
3. Industrial-Strength Natural Language Processing (2019). https://spacy.io/. Accessed 29 Mar 2019
4. Krisdika (2019). http://www.krisdika.go.th/. Accessed 29 Mar 2019
5. Lembaran Negara (2019). http://ditjenpp.kemenkumham.go.id/kerja/lnnew.php. Accessed 10 July 2019
6. Mundaneum (2019). https://github.com/jackrusher/mundaneum. Accessed 22 Sept 2019
7. Nokogiri (2019). https://nokogiri.org/. Accessed 29 Mar 2019
8. Official Gazette (2019). http://vietnamlawmagazine.vn/gazette.html. Accessed 29 Mar 2019
9. Official Portal Attorney General's Chambers of Malaysia (2019). http://www.agc.gov.my/agcportal. Accessed 29 Mar 2019
10. Optical Character Recognition (OCR): Tutorial—cloud functions document—Google cloud (2019). https://cloud.google.com/functions/docs/tutorials/ocr. Accessed 29 Mar 2019
11. Singapore statutes online (2019). https://sso.agc.gov.sg/. Accessed 29 Mar 2019
12. Socialist Republic of Vietnam Government Portal (2019). http://congbao.chinhphu.vn/cong-bao-nam-2019. Accessed 28 Sept 2019
13. Thai word segmentation library in Rust (2019). https://github.com/veer66/chamkho. Accessed 29 Mar 2019
14. The national assembly of the Lao people's democratic republic (2019). http://www.na.gov.la/. Accessed 29 Mar 2019
15. The official Gazette of the Republic of the Philipines (2019). https://www.officialgazette.gov.ph. Accessed 29 Mar 2019
16. Bojanowski, P., Grave, E., Joulin, A., Mikolov, T.: Enriching word vectors with subword information. Trans. Assoc. Comput. Linguist. **5**, 135–146 (2017)
17. Dev, S.: Slimerjs (2019). https://slimerjs.org/. Accessed 12 July 2019
18. Eberle, E.J.: The method and role of comparative law. Wash. Univ. Glob. Stud. Law Rev. **8**, 451 (2009)
19. Harris, S., Seaborne, A., Prud'hommeaux, E.: SPARQL 1.1 query language. W3C recommendation (2013). Accessed 23 Sept 2019
20. Iacobacci, I., Pilehvar, M.T., Navigli, R.: Embeddings for word sense disambiguation: an evaluation study. In: Proceedings of the 54th Annual Meeting of the Association for Computational Linguistics (Volume 1: Long Papers), pp. 897–907 (2016)
21. Koehn, P., et al.: Moses: open source toolkit for statistical machine translation. In: Proceedings of the 45th Annual Meeting of the Association for Computational Linguistics Companion Volume Proceedings of the Demo and Poster Sessions, pp. 177–180 (2007)
22. Mozilla: Mozilla Firefox (2019). https://www.mozilla.org/th/. Accessed 12 July 2019
23. Tiedemann, J.: Parallel data, tools and interfaces in OPUS. In: LREC, vol. 2012, pp. 2214–2218 (2012)

Author Index

Printed in the United States
By Bookmasters